SpringerBriefs in Speech Technology

Studies in Speech Signal Processing, Natural Language Understanding, and Machine Learning

Series Editor

Amy Neustein, Fort Lee, NJ, USA

SpringerBriefs present concise summaries of cutting-edge research and practical applications across a wide spectrum of fields. Featuring compact volumes of 50 to 125 pages, the series covers a range of content from professional to academic. Typical topics might include:

- A timely report of state-of-the-art analytical techniques
- A bridge between new research results, as published in journal articles, and a contextual literature review
- A snapshot of a hot or emerging topic
- An in-depth case study or clinical example
- A presentation of core concepts that students must understand in order to make independent contributions

Briefs are characterized by fast, global electronic dissemination, standard publishing contracts, standardized manuscript preparation and formatting guidelines, and expedited production schedules.

The goal of the **SpringerBriefs in Speech Technology** series is to serve as an important reference guide for speech developers, system designers, speech engineers and other professionals in academia, government and the private sector. To accomplish this task, the series will showcase the latest findings in speech technology, ranging from a comparative analysis of contemporary methods of speech parameterization to recent advances in commercial deployment of spoken dialog systems.

More information about this series at http://www.springer.com/series/10043

Prajna Kunche • N. Manikanthababu

Fractional Fourier Transform Techniques for Speech Enhancement

Springer

Prajna Kunche
Indira Gandhi Centre for Atomic
Research
Kalpakkam, Tamil Nadu, India

N. Manikanthababu
Indira Gandhi Centre for Atomic Research
Kalpakkam, Tamil Nadu, India

ISSN 2191-737X   ISSN 2191-7388   (electronic)
SpringerBriefs in Speech Technology
ISBN 978-3-030-42745-0   ISBN 978-3-030-42746-7   (eBook)
https://doi.org/10.1007/978-3-030-42746-7

This Springer imprint is published by the registered company Springer Nature Switzerland AG
The registered company address is: Gewerbestrasse 11, 6330 Cham, Switzerland

Preface

This book expounds the use of fractional Fourier transform techniques for speech enhancement applications. It could serve as a resource guide for the engineers, scientists, and academic researchers who are working in this area.

Speech enhancement involves the processing of degraded speech. It has been a challenging problem over the decades and yet is an active area of research due to the complexities involved in the highly non-stationary signal processing. In general, the processing is in temporal or spectral domains. Many expedient techniques have been proposed to non-stationary signal processing such as short-time Fourier transform (STFT), wavelet transform (WT), and fractional Fourier transform (FrFT). Among all the transformation techniques, FrFT had been proved to be a perfect time-frequency analysis tool in many signal processing applications. With this context, this book framed to explain the speech enhancement in FrFT domain and also explores the use of different FrFT algorithms in both single-channel and multi-channel enhancement systems.

The primary goal of the book is to provide a rigorous introduction to the major concepts of the fractional Fourier transform: What if FrFT? Why FrFT is advantageous for time-frequency analysis? What are the properties of FrFT? What are the numerous applications of FrFT?

The secondary goal of this book is to present new FrFT algorithms which were proposed for speech enhancement. All the new approaches are presented with the apposite results and analysis.

This book is organized into six chapters. The earlier chapters give a comprehensive development of fractional Fourier transform, and the later chapters present the generalization and the application of FrFT in speech enhancement. Special focus is given on the detailed explanation of fractional cosine transform and fractional sine transform.

This book is arranged in a progressive order, starting with basic definitions, moving to a thorough discussion on the enactment of the algorithms. This could help the readers to understand and also to develop new algorithms easily.

Some of the unique features of the book include:

- A comprehensive analysis of fractional Fourier transform techniques in speech enhancement applications
- New approaches for speech enhancement using FrFT
- Attempts are made to elucidate the theory of different FrFT methods in detail.
- The future scope of research in this emerging area

Kalpakkam, India   Prajna Kunche
Kalpakkam, India   N. Manikanthababu

Contents

Chapter 1
Introduction

1.1 Speech Enhancement Algorithms

Speech enhancement refers to improving the speech quality by suppressing or estimating the background noise. The speech enhancement algorithms are classified based on the number of microphones used such as single channel systems, multichannel systems or dual channel systems (adaptive noise cancellation), and multisensory beamforming. Speech enhancement algorithms are also classified as time domain, frequency domain, and time-frequency representation methods. Typically, many algorithms for processing audio signals do not work on raw time domain audio signals, but on time-frequency representations. A basic audio/speech signal encodes the amplitude of the sound as a function of time. In frequency analysis, the Fourier spectrum gives the signal information as the function of frequency but it cannot reflect the variations on time plane. In time-frequency processing of audio signal the amplitude of the sound signal is represented in both time and frequency domain and is able to provide the combined information of temporal and spectral characteristics. The various speech enhancement algorithms that have been developed in time and transform domain algorithms are shown in Fig. 1.1. The discrete Fourier transform (DFT), Karhunen–Loeve transform (KLT), discrete cosine transform (DCT), discrete wavelet transform (DWT), and the fractional Fourier transform (FrFT) are different transforms used for speech enhancement algorithms. Each of the algorithms is explained in detail in the following sections.

P. Kunche, N. Manikanthababu, *Fractional Fourier Transform Techniques for Speech Enhancement*, SpringerBriefs in Speech Technology, https://doi.org/10.1007/978-3-030-42746-7_1

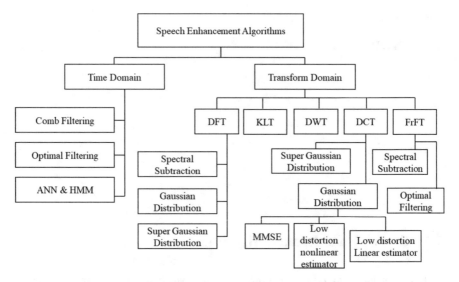

Fig. 1.1 Classification of speech enhancement algorithms in time and transform domains

1.2   Time Domain Enhancement Methods

1.2.1   Comb Filtering

Some researchers have studied the adaptive comb filter for speech signal noise reduction applications in time domain. Shields et al. studied comb filtering operations in which the harmonics of the speech only passes through the filter. In general, the noise signals possess energy in the frequencies in between the speech harmonics. Hence this method can give effective results when the information of fundamental frequency is available and also there is no loss in the periodicity of speech. As per the studies of Frazier, Shields method suffers with distortion of speech because of the non-stationary behavior of speech signal. To overcome this, Frazier proposed another method in which the comb filter adjusts itself to the fundamental frequency. Perlmutter et al. utilized Frazier's algorithm to perform intelligibility tests with interference consisting of the speech and she reported that there is loss in the intelligibility for some range of signal to noise ratios (SNR). Later, Lim et al. (1978) reported that Frazier's algorithm cannot increase the intelligibility of enhanced speech independent of the signal's SNR.

The principle of adaptive comb filter could be described by taking its unit sample response for one pitch period:

$$h(n) = \sum_{k=-L}^{L} a_k \cdot \delta(n - N_k)$$

where $h(n)$ is a unit sample response and $\delta(n)$ is a unit sample function and the length of the filter is $2L + 1$ pitch periods, a_k is the filter coefficients that hold $\sum_{k=-L}^{L} a_k = 1$, and N_k is given by

$$N_k = -\sum_{l=k}^{-1} T_l \quad \text{for } k < 0 \tag{1.1}$$

$$= \quad 0 \quad \text{for } k = 0 \tag{1.2}$$

$$= \sum_{l=0}^{k-1} T_l \quad \text{for } k > 0 \tag{1.3}$$

Here, T_k corresponds to the particular pitch period that consists the point of speech waveform which is multiplied by the filter coefficient a_k. Depending on the pitch information of speech waveform, N_k is updated for every pitch period.

1.2.2 Subspace Optimal Filtering

Reduced rank or subspace methods are the class of time domain approaches. The concept of reduced rank is first introduced in the field of estimation. Later it is applied for subspace noise reduction. In this method, first the noise is estimated using the singular value decomposition (SVD) of the noisy data matrix and then the clean speech is estimated from the remaining subspace. Researchers also introduced other techniques such as eigenvalue decomposition (EVD), truncated SVD (QSVD), and triangular decompositions for subspace approach. Zhang et al. (2016) proposed single channel noise reduction in time domain by using semi-orthogonal transformations and reduced-rank filtering. In this method authors transformed the observation signal vector sequence, through a semi-orthogonal matrix, into a sequence of transformed signal vectors with a reduced dimension. Second, a reduced-rank filter is applied to get an estimate of the clean speech in the transformed domain. Third, the estimate of the clean speech in the time domain is obtained by an inverse semi-orthogonal transformation.

The signal model considered for these approaches is as follows:

$$y(k) = x(k) + v(k) \tag{1.4}$$

where $x(k)$ is the desired speech signal, $y(k)$ is the noisy observation and k is the discrete time index of the zero-mean from the noisy observation, and $v(k)$ is the additive noise. Let us consider that all signals are zero-mean and broad band.

This signal model can be represented in a vector form as shown below:

$$y(k) = x(k) + v(k) \tag{1.5}$$

where $y(k) \triangleq [y(k) \quad y(k-1)...y(k-L+1)]^T$ is a vector of length L and the subscript T refers to the transpose of the matrix. The vectors $x(k)$ and $v(k)$ are defined in similar way to $y(k)$.

Assume that $x(k)$ and $v(k)$ are the uncorrelated then the correlation matrix of noisy signal can be defined as

$$R_y \triangleq E\left[y(k)y^T(k)\right] = R_x + R_v \tag{1.6}$$

where the operator $E[\cdot]$ denotes the mathematical expectation and R_x and R_v are the correlation matrices of $x(k)$ and $v(k)$, respectively. R_x and R_v are defined as

$$R_x \triangleq E\left[x(k)x^T(k)\right] \tag{1.7}$$

$$R_v \triangleq E\left[v(k)v^T(k)\right] \tag{1.8}$$

The noise in the speech signal will be reduced by proper estimation of vector $x(k)$ from noisy observation $y(k)$.

1.2.3 Artificial Neural Network for Speech Enhancement

The explicit, nonlinear time domain filter is the simplest way of the neural network for speech enhancement (Ram and Mohanty 2018; Wang and Wang 2015; Yu et al. 2019). The input corrupted noisy speech signals and target clean speech signals are fed into the NN for training. With sufficient training, the NN will be able to remove background noise and thus improve the quality of speech during the testing stage. A multi-layered network is used in this approach to map the windowed portion of the noisy speech to an approximation of its clean voice. The number of inputs is dependent on the sampling rate of the speech signal. The number of outputs is generally equal to number of inputs. First an artificial noise is added to the clean speech to generate noisy signal, to train the network. The clean speech is considered as the target that is time aligned with inputs. Any type of training approaches can be employed and usually a standard backpropagation.

The input window is moving across the noisy speech signal to present the data. The increment in the window shift is represented as L at each step, between 1 and the window length M. The estimation window slides without overlap when $L = M$. For increments $L < M$, the resultant overlapping windows provide redundant estimates.

Deep learning is an emerging field in current research. A deep neural network (DNN) consists of the convolutionary neural network (CNN) and the deep belief network (DBN) as the deep learning models.

Yu et al. (2019) and others proposed deep neural network Kalman filtering based time domain speech enhancement algorithm. The overall system in this approach is in two stages: training stage and enhancement stage. In training stage, a DNN is trained to learn the mapping from the line spectrum frequencies (LSF) of noisy speech to the clean speech LSFs. In the enhancement stage, a DNN based Kalman filter estimated parameters are applied to the noisy speech to get the enhanced speech.

In training stage the LPCs are calculated for both noisy and clean speech databases and then they are converted into LSFs to improve the stability of the training stage. For the DNN, the noisy LSFs are considered as input features and the clean LSFs are taken as output targets. The mapping of noisy LSFs into LSFs of clean ones is a highly nonlinear regression function. To perform supervised learning, authors used a feed forward neural network with many nonlinear units. For training of DNN, back propagation with the MMSE based cost function is utilized. In the enhancement stage, the DNN-based Kalman filter algorithm is described as:

Step 1:   Estimating clean LSFs from noisy LSFs with the DNN.
Step 2:   Converting the LSFs to LPCs to form the state transition matrix.
Step 3:   Computing the covariance matrix of the measured noise and driving noise.
Step 4:   Perform Kalman filter to get the filtered estimate of the speech state vector.
Step 5:   Finally, obtain the speech estimate using the filtered estimate of the speech state vector.

1.3   Transform Domain Speech Enhancement Algorithms

1.3.1   Discrete Fourier Transform

In Gaussian distribution method, the MMSE-STSA (Ephraim and Malah 1984) is derived based on modeling the speech and noise spectral components as statistically independent Gaussian random variables. The STSA estimation problem formulated in this work is that of estimating the modulus of each complex Fourier expansion coefficient of the speech signal in a given analysis frame from the noisy speech in that frame. Using this approach, the gain can be calculated using a priori SNR and a posteriori SNR. The a priori SNR is defined as

$$\varepsilon(n.k) = \frac{\lambda_s(n, k)}{\lambda_d(n, k)} \tag{1.9}$$

The posteriori SNR is given as

$$\gamma(n, k) = \frac{|X(n, k)|}{\lambda_d(n, k)} \tag{1.10}$$

The MMSE-STSA estimator gain function is defined as

$$G_{\text{MMSE–STSA}}(n, k) = \frac{\sqrt{\pi}}{2} \frac{\sqrt{v(n, k)}}{\gamma(n.k)} \exp\left(\frac{-v(n, k)}{2}\right)$$
$$\times \left((1 + v(n, k))I_0\left(\frac{v(n, k)}{2}\right) + v(n, k)I_1\left(\frac{v(n, k)}{2}\right)\right) \tag{1.11}$$

Here, I_0 and I_1 are the modified Bessel functions of zero and first order, respectively, and $v(n, k)$ is defined as

$$v(n, k) = \frac{\varepsilon(n.k)}{\varepsilon(n.k) + 1}\gamma(n.k) \tag{1.12}$$

Ephraim and Malah (1985) derived another type of MMSE-STSA estimator that minimizes the mean square error of the log spectra namely MMSE-log-STSA estimator. The gain function of the MMSE-log-STSA estimator is defined as

$$G_{\text{MMSE–log–STSA}}(n, k) = \frac{\varepsilon(n.k)}{\varepsilon(n.k) + 1} \exp\left\{\frac{1}{2}\int_{v(n,k)}^{\infty} \frac{e^{-t}}{t} dt\right\} \tag{1.13}$$

1.3.2   Wavelet Transform

Hu et al. (Ephraim and Van Trees 1995) proposed the approach of wavelet thresholding the multitaper spectrum for speech enhancement. In this work, they used low variance spectrum estimators based on wavelet thresholding and derived a short time spectral amplitude with wavelet thresholding multitaper spectra. In speech enhancement, the direct spectrum estimation based on Hamming window is the common estimation method for power spectrum. The limitation of this windowing method is it cannot reduce the variance of the spectral estimate. The concept behind the multitaper spectrum estimator is to minimize this variation by measuring a small number of direct spectrum estimators each with a different taper (window) and then averaging the spectral estimates. If the L tapers are chosen to be orthogonal in pairs and correctly designed to avoid leakage, the resulting multitaper spectral estimator in terms of reduced bias and variance will be superior to the periodogram. At best, the multitaper prediction variance will be lower than the variance of each spectral estimation by the factor 1/L.

The multitaper spectrum estimate is defined as

$$\widehat{S}^{mt}(\omega) = \frac{1}{L}\sum_{k=0}^{L-1}\widehat{S}_k^{mt}(\omega) \tag{1.14}$$

$$\text{where } \widehat{S}_k^{mt}(\omega) = \left|\sum_{m=0}^{N-1}a_k(m)x(m)e^{-j\omega m}\right|^2$$

where N is the data length and a_k is the kth data taper used for the spectral estimate $\widehat{S}_k^{mt}(\cdot)$, which is also called the kth eigen spectrum. The taper sequences are the sine tapers given as

$$a_k(m) = \sqrt{\frac{2}{N+1}}\sin\frac{\pi k(m+1)}{N+1}, \qquad m = 0, 1, \dots N-1 \tag{1.15}$$

Wavelet thresholding can be used to refine the spectral estimate and produce a smooth estimate of the logarithmic of the spectrum. The refining of multitaper spectrum by wavelet thresholding is as follows:

Step 1:   Obtain the multitaper spectrum using the Eqs. (1.14) and (1.15) and determine $Z(\omega)$ using following equation:

$$Z(\omega) = \log\widehat{S}^{mt}(\omega) - \varnothing(L) + \log L \tag{1.16}$$

Step 2:   Apply a discrete wavelet transform (DWT) to $Z(\omega)$ to obtain the empirical DWT coefficients $Z_{j,\,k}$ for each level j.

Step 3:   Apply thresholding procedure to $Z_{j,\,k}$.

Step 4:   Apply the inverse DWT to the thresholded wavelet coefficients to get the refined log spectra.

The STSA estimator with wavelet thresholding method is a ratio of multitaper spectra $\widehat{S}_x^{mt}(\omega)/\widehat{S}_n^{mt}(\omega)$ where $\widehat{S}_x^{mt}(\omega)$ is the clean speech and $\widehat{S}_n^{mt}(\omega)$ is the estimate of the noise spectrum and the wavelet threshold of the log of the ratio of the two spectrum are formed to get estimate of $\gamma_{prio}(k)$.

The log of a priori SNR estimate with multitaper spectra is modeled as the true log a priori SNR plus a Gaussian distributed noise $\varepsilon(k)$, i.e.,

$$\log\gamma_{prio}^{mt}(k) = \log\gamma_{prio}(k) + \varepsilon(k) \tag{1.17}$$

The wavelet based denoising technique can be used to eliminate $\varepsilon(k)$.

Another approach is based on the assumption that it is possible to obtain a good estimate of a priori SNR using a good low variance spectral approximation of $\widehat{S}_x(\omega)$

and $\widehat{S}_n(\omega)$. In addition to the wavelet threshold estimate of $\widehat{S}_n(\omega)$, the refined spectrum $\widehat{S}_x(\omega)$ is used to obtain a better estimate of the a priori SNR.

1.3.3 KL Transform

Karhunen–Loeve transform (KLT) (Ephraim and Van Trees 1995) is applied to noisy speech enhancement by Y. Ephraim in 1995. Each of the estimators using KL transform is described as below:

(a) LMMSE Estimation:
  Considering the fact that some signal eigenvalues must be zero, in this method the number of zero eigenvalues are eliminated first and then the corresponding KLT components must be eliminated, otherwise the negative estimates of signal eigenvalues give flaws in the estimation.
(b) LS Estimation:
  The LS estimation of y given x is obtained from

$$\min_{\{y=Vs\}} \|z - y\|^2 \tag{1.18}$$

where V is assumed as known value. This estimator is formulated as

$$\widehat{y}_{ls} = V\left(V^\# V\right)^{-1} V^\# z \tag{1.19}$$

$$\triangleq P_V z \tag{1.20}$$

$$P_V = U_1\left(U_1^\# U_1\right)^{-1} U_1^\# = U_1 U_1^\# \tag{1.21}$$

An LS estimation of V can be obtained from the joint LS estimations of s and V which is given by

$$\min_{\{s,\,V\}} \|z - Vs\|^2 \tag{1.22}$$

Then we get

$$\widehat{s}_{ls} = \left(V^\# V\right)^{-1} V^\# z \tag{1.23}$$

When substituting $\widehat{s}_{ls}$ we get

$$\max_V \operatorname{tr}\left\{P_V z z^\#\right\} \tag{1.24}$$

(c) Spectral subtraction estimation:

In discrete Fourier transform (DFT) based spectral subtraction approach (Boll 1979), first the noisy speech is transformed into frequency domain. The spectral components whose estimated variance is smaller than the spectral components of noise estimates are nullified. A gain function modifies the remaining spectral components and then inverse DFT (IDFT) is applied to transform back the signal into time domain. If we presume the DFT will have desired properties of KLT, then the spectral subtraction and noise estimation are explained as follows:

Assume $D^{\#}$ represents the $K \times K$ matrix of DFT and $D^{\#}z$ denotes the vector of spectral components of the noisy signal. The noise signal spectral components are represented by vector $D^{\#}w$. Let us assume that only $M < K$ spectral components of vector $D^{\#}z$ are having their variance greater than the variance of the corresponding component of $D^{\#}w$. If the columns of D resulting in these M components are arranged in a matrix of $K \times M$ referred to as D_1, and $D = [D_1, D_2]$, then the spectral subtraction estimator can defined as

$$\widehat{y}_{\text{sps}} = \frac{1}{K} D \begin{bmatrix} G_{\text{sps}} & 0 \\ 0 & 0 \end{bmatrix} D^{\#}z \qquad (1.25)$$

$$= \frac{1}{K} D_1 G_{\text{sps}} D_1^{\#}z \qquad (1.26)$$

here, G_{sps} is a diagonal gain matrix of size $M \times M$. If the gain function of Wiener filter is considered, then the mth diagonal element of G_{sps} is defined by

$$g_{\text{sps}} = \frac{\widehat{f}_z(m) - \widehat{f}_w(m)}{\widehat{f}_z(m)} \qquad (1.27)$$

where $\widehat{f}_z(m) = E\left\{\left|(D^{\#}z)_m\right|^2\right\}/K$ refers to the mth spectral component of the noisy signal and $\widehat{f}_w(m) = E\left\{\left|(D^{\#}w)_m\right|^2\right\}/K$ represents the variance of the mth spectral component of the noise signal. Here $\widehat{f}_z(m)$ and $\widehat{f}_w(m)$ are the estimates of the power spectral densities of the noisy signal and noise, respectively.

1.3.4   Discrete Cosine Transform

Noisy speech enhancement using discrete cosine transform (DCT) has also been implemented by considering the advantages of DCT over DFT (Soon et al. 1998). These advantages of DCT can be summarized as follows:

- This has a substantially higher energy compaction capability;
- This is a real transformation without the knowledge of the phase;
- This provides a higher resolution to evaluate the coefficients of transformation in the same window size.

$$y(t) = x(t) + n(t) \tag{1.28}$$

$Y(k)$, $X(k)$, and $N(k)$ are the transformed signals where k represents the position of the signal in transform domain. Minimum mean square error (MMSE) estimated amplitude $\widehat{X}(k)$ can be obtained from $Y(k)$ as follows:

$$\widehat{X}(k) = E\{X(k)|Y(k)\} \tag{1.29}$$

where $E\{\cdot\}$ denotes the expectation operator and Eq. (1.29) can be rewritten, using Bayes' theorem, as

$$\widehat{X}(k) = \frac{\int_{-\infty}^{\infty} a_k p\{Y(k)|a_k\} p\{a_k\} da_k}{\int_{-\infty}^{\infty} p\{Y(k)|a_k\} p\{a_k\} da_k} \tag{1.30}$$

Under the Gaussian distribution assumptions, $p\{Y(k)|a_k\}$ and a_k are given by the following equations:

$$p\{Y(k)|a_k\} = \frac{1}{\sqrt{2\pi\lambda_n(k)}} \exp\left\{ -\frac{(Y(k) - a_k)^2}{2\lambda_n(k)} \right\} \tag{1.31}$$

$$p\{a_k\} = \frac{1}{\sqrt{2\pi\lambda_x(k)}} \exp\left\{ -\frac{(-a_k)^2}{2\lambda_x(k)} \right\} \tag{1.32}$$

where $\lambda_x(k) = E\{|X(k)|^2\}$ and $\lambda_n(k) = E\{|N(k)|^2\}$

Substituting Eq. (1.32) and Eq. (1.31) in Eq. (1.30), we get

$$\widehat{X}(k) = \frac{\xi(k)}{\xi(k) + 1} Y(k) \tag{1.33}$$

where

$$\xi(k) = \frac{\lambda_x(k)}{\lambda_n(k)} \tag{1.34}$$

The estimate $\widehat{\lambda}_x$ for λ_x is given by the following equation:

$$\widehat{\lambda}_x(k) = \alpha\widehat{\lambda}_x(k)_p + (1 - \alpha) \max\left\{ Y(k)^2 - \lambda_n(k), 0 \right\} \tag{1.35}$$

Zou et al. (2007) derived MMSE estimator based on the DCT coefficients of the speech. The DCT coefficients of clean speech are modeled by a Laplacian or a Gamma distribution and the DCT coefficients of the noise are Gaussian distributed. They also derived the MMSE estimator with speech presence uncertainty. Each of the algorithms is modeled as follows:

In the general statistical model of DCT based speech enhancement, the noisy speech signal $y(t) = x(t) + n(t)$ is sampled and segmented as frames by applying a window function (e.g., normalized Hann window). Then the DCT coefficients of frame l and frequency bin k are determined by using the following equation:

$$Y(k, l) = X(k, l) + N(k, l) \tag{1.36}$$

$$= a(k) \sum_{n=0}^{L-1} y(lR + 1)h(n) \cos \left[\frac{\pi(2N + 1)k}{2L} \right] \tag{1.37}$$

where $a(0) = \sqrt{\frac{1}{N}}$, $a(k) = \sqrt{\frac{2}{N}}$, $1 \leq k \leq L - 1$ and L is the length of the window. The window is moved by R samples to determine the next DCT. The distribution of noise coefficients is modeled by the Gaussian assumption given by

$$P(D) = \frac{1}{2\pi\sigma_d} \exp\left(-\frac{D^2}{2\sigma_d^2}\right) \tag{1.38}$$

where σ_d^2 is the noise variance.

For the speech coefficients the conventional method uses the Gaussian distribution and some researchers proved that the PDF of the DCT coefficients of speech are better approximated with a Laplacian or Gamma models. Each model is defined as follows:

1. Gaussian distribution of speech model:

$$P(X) = \frac{1}{2\pi\sigma_x} \exp\left(-\frac{X^2}{2\sigma_x^2}\right) \tag{1.39}$$

2. Laplacian distribution of speech model:

$$P(X) = \frac{1}{2\alpha} \exp\left(-\frac{|X|}{\alpha}\right) \tag{1.40}$$

3. Gamma distribution of speech model:

$$P(X) = \frac{\sqrt[4]{3}}{2\sqrt{2\pi\sigma_x}} |X|^{-\frac{1}{2}} \exp\left(-\frac{\sqrt{3}|X|}{2\sigma_x}\right) \tag{1.41}$$

where σ_x^2 is the variance of speech, $\alpha = E\{|X|\}$ is the Laplacian factor, and $E\{\cdot\}$ represents the expectation operator.

The MMSE estimator of X is given by

$$\widehat{X} = E\{X|Y\} = \int_{-\infty}^{\infty} X f_{X|Y}(X|Y) dX \tag{1.42}$$

$$= \frac{\int_{-\infty}^{\infty} X f_{X,Y}(X,Y) dX}{\int_{-\infty}^{\infty} f_{X,Y}(X,Y) dX} \tag{1.43}$$

Gaussian Based MMSE Estimator

Let us assume the speech coefficients and noise coefficients PDFs are Gaussian, $f_{X,Y}(X,Y)$ is given by

$$f_{X,Y}(X,Y) = \frac{1}{\sqrt{2\pi}\sigma_x} \exp\left(-\frac{X^2}{2\sigma_x^2}\right) \frac{1}{\sqrt{2\pi}\sigma_d} \exp\left(-\frac{(Y-X)^2}{2\sigma_d^2}\right) \tag{1.44}$$

On substituting Eq. (1.44) into Eq. (1.42), the following equation can be obtained:

$$\widehat{X} = E\{X|Y\} = \frac{\sigma_x^2}{\sigma_x^2 + \sigma_d^2} Y \tag{1.45}$$

$$= \frac{\xi}{1+\xi} Y \tag{1.46}$$

here, $\xi = \sigma_x^2/\sigma_d^2$, represents a priori signal to noise ratio and is estimated using decision direct method.

$$\widehat{\xi}(k,l) = \alpha_\xi \frac{\left|\widehat{X}(k,l-1)\right|^2}{\sigma_d^2(k,l-1)} + (1-\alpha_\xi) \max\{\psi(k,l)-1,0\} \tag{1.47}$$

where α_ξ is a smoothing factor in the range $0 \leq \alpha_\xi \leq 1$ and $\psi = |Y|/\sigma_d^2$ is a posteriori SNR.

Laplacian Based MMSE Estimator

In this method, the speech and noise coefficients PDFs are assumed as Laplacian, then $f_{X,Y}(X,Y)$ is given as

$$f_{X,Y}(X,Y) = \frac{1}{2\alpha} \exp\left(-\frac{|X|}{\alpha}\right) \cdot \frac{1}{\sqrt{2\pi}\sigma_d} \exp\left(-\frac{(Y-X)^2}{2\sigma_d^2}\right) \tag{1.48}$$

On substituting Eq. (1.48) into Eq. (1.42), we get

$$\widehat{X} = E\{X|Y\} \tag{1.49}$$

$$= \sigma_d \left[\left(\psi + \frac{1}{\xi} \right) \exp \left(\left(\psi + \frac{1}{\xi} \right)^2 / 2 \right) . \right.$$

$$\mathrm{erfc} \left[\left[\left(\psi + \frac{1}{\xi} \right) / \sqrt{2} \right] - \left(-\psi + \frac{1}{\xi} \right) \right] . \exp \left[\left(-\psi + \frac{1}{\xi} \right)^2 / 2 \right].$$

$$\mathrm{erfc} \left[\left[\left(-\psi + \frac{1}{\xi} \right) / \sqrt{2} \right] \right] / \exp \left[\left(-\psi + \frac{1}{\xi} \right)^2 / 2 \right] \mathrm{erfc} \left[\left[\left(-\psi + \frac{1}{\xi} \right) / \sqrt{2} \right] \right]. \quad (1.50)$$

$$\mathrm{erfc}(x) = \frac{2}{\sqrt{\pi}} \int_x^{+\infty} \exp \left(-t^2 \right) dt \quad (1.51)$$

$$\widehat{\xi}(k,l) = \alpha_\xi \frac{\left| \widehat{X}(k,l-1) \right|^2}{\sigma_d^2(k,l-1)} + (1 - \alpha_\xi) \max \{ \psi(k,l) - 1, 0 \} \quad (1.52)$$

Gamma Based MMSE Estimator

Let us assume that speech coefficients in PDF is a Gamma, $f_{X,Y}(X,Y)$ can be written as

$$f_{X,Y}(X,Y) = \frac{\sqrt[4]{3}}{2\sqrt{2\pi}\sigma_x} |X|^{-\frac{1}{2}} \exp \left(-\frac{|X|}{2\sigma_x} \right) . \frac{1}{\sqrt{2\pi}\sigma_d} \exp \left(-\frac{|Y-X|^2}{2\sigma_d^2} \right) \quad (1.53)$$

$$\widehat{X} = E\{X|Y\} \quad (1.54)$$

$$= \sigma_d \frac{\Gamma(0.5)}{\Gamma(0.5)} \left[\exp \left(\left(\frac{\sqrt{3}}{2\sqrt{\xi}} - \psi \right)^2 / 4 \right) D_{-1.5} \left(\frac{\sqrt{3}}{2\sqrt{\xi}} - \psi \right) - \exp \left(\frac{\left(\frac{\sqrt{3}}{2\sqrt{\xi}} + \psi \right)^2}{4} \right) D_{-1.5} \left(\frac{\sqrt{3}}{2\sqrt{\xi}} + \psi \right) \right] /$$

$$\left[\exp \left(\left(\frac{\sqrt{3}}{2\sqrt{\xi}} - \psi \right)^2 / 4 \right) D_{-0.5} \left(\frac{\sqrt{3}}{2\sqrt{\xi}} - \psi \right) + \exp \left(\left(\frac{\sqrt{3}}{2\sqrt{\xi}} + \psi \right)^2 / 4 \right) D_{-0.5} \left(\frac{\sqrt{3}}{2\sqrt{\xi}} + \psi \right) \right] \quad (1.55)$$

$$\xi = \frac{\sigma_x^2}{\sigma_d^2} \text{ and } \psi = Y/\sigma_d.$$

MMSE Estimator with Speech Presence Uncertainty

Let H_1 and H_0 be the two states that represent the speech presence and speech absence, respectively. The new MMSE estimator is given by

$$\widehat{X} = E\{X|Y, H_1\} P(H_1|Y) \quad (1.56)$$

On the application of Bayes rule for the conditional speech presence, we get

$$P(H_1|Y) = \frac{\Lambda(Y,q)}{1 + \Lambda(Y,q)} \tag{1.57}$$

$$\Lambda(Y,q) = \frac{1-q}{q}\frac{P(Y|H_1)}{P(Y|H_0)} \tag{1.58}$$

here $q = P(H_0)$ is the a priori probability for speech absence. If we consider that noise coefficients PDF is Gaussian, $P(Y|H_0)$ is in the following form:

$$P(Y|H_0) = \frac{1}{\sqrt{2\pi}\sigma_d}\exp\left(-\frac{|Y|}{2\sigma_d^2}\right) \tag{1.59}$$

$$= \frac{1}{\sqrt{2\pi}\sigma_d}\exp\left(-\frac{\psi^2}{2}\right) \tag{1.60}$$

Laplacian Speech Model
Under the assumption that the speech coefficients PDF is a Laplacian, $P(Y|H1)$ will have the form

$$P(Y|H_1) = \int_{-\infty}^{\infty} f_{X,Y}(X,Y)dX \tag{1.61}$$

$$= \frac{1}{4\alpha}\exp\left(\frac{1}{2\xi^2}\right)$$

$$\times \left\{\exp\left(\frac{\psi}{\xi}\right)\text{erfc}\left[\left(\psi + \frac{1}{\xi}\right)/\sqrt{2}\right] + \exp\left(-\frac{\psi}{\xi}\right)\text{erfc}\left[\left(-\psi + \frac{1}{\xi}\right)/\sqrt{2}\right]\right\} \tag{1.62}$$

So the generalized likelihood ratio $\Lambda(Y,q)$ has the form

$$\Lambda(Y,q) = \frac{1-q}{q}\frac{\sqrt{2\pi}}{4\xi}\left\{\exp\left[\frac{\left(\psi + \frac{1}{\xi}\right)^2}{2}\right].\text{erfc}\left[\frac{\left(\psi + \frac{1}{\xi}\right)}{\sqrt{2}}\right]\right.$$

$$\left. + \exp\left[\frac{\left(-\psi + \frac{1}{\xi}\right)^2}{2}\right].\text{erfc}\left[\frac{\left(-\psi + \frac{1}{\xi}\right)}{\sqrt{2}}\right]\right\} \tag{1.63}$$

Gamma Speech Model
Under the assumption that the speech coefficients PDF is a Gamma, $P(Y|H1)$ will have the form

$$P(Y|H_1) = \int_{-\infty}^{\infty} f_{X,Y}(X,Y)dX \tag{1.64}$$

$$
= \frac{\exp\left(-\frac{Y}{2\sigma_d^2}\right)}{\sqrt{2\pi}\sigma_d} \frac{\sqrt[4]{3}\,\sqrt{\sigma_d}}{2\sqrt{2\pi}\sigma_x} \Gamma(0.5)
$$

$$
\left\{ \exp\left(\left(\frac{\sqrt{3}}{2\sqrt{\xi}} - \psi\right)^2/4\right) D_{-0.5}\left(\frac{\sqrt{3}}{2\sqrt{\xi}} - \psi\right) + \exp\left(\left(\frac{\sqrt{3}}{2\sqrt{\xi}} - \psi\right)^2/4\right) D_{-0.5}\left(\frac{\sqrt{3}}{2\sqrt{\xi}} - \psi\right) \right\}
$$

$$(1.65)$$

So the generalized likelihood ratio $\bigwedge(Y,q)$ has the form

$$
\bigwedge(Y,q) = \frac{1-q}{q} \frac{\sqrt[4]{3}\,\sqrt{\sigma_d}}{2\sqrt{2\pi}\sigma_x} \Gamma(0.5)
$$

$$
\left\{ \exp\left(\left(\frac{\sqrt{3}}{2\sqrt{\xi}} - \psi\right)^2/4\right) D_{-0.5}\left(\frac{\sqrt{3}}{2\sqrt{\xi}} - \psi\right) + \exp\left(\left(\frac{\sqrt{3}}{2\sqrt{\xi}} - \psi\right)^2/4\right) D_{-0.5}\left(\frac{\sqrt{3}}{2\sqrt{\xi}} - \psi\right) \right\}
$$

$$(1.66)$$

MMSE Super Gaussian

This MMSE estimator is the conditional mean of s with $f_{s \mid v}(s \mid v)$ as the PDF (Zou and Zhang 2007). The MMSE estimator of the clean speech component is given as a nonlinear function of three inputs: (1) noisy speech component, (2) noise variance σ^2, and (3) speech Laplacian factor a

$$
\hat{s} \triangleq E\{s \mid v\} = \int_{-\infty}^{\infty} s f_{s \mid v}(s \mid v) ds \tag{1.67}
$$

$$
= a e^{\frac{\psi}{2}} \left[\frac{(\psi + \xi) e^{\xi} \mathrm{erfc}\left(\frac{\psi+\xi}{\sqrt{2\psi}}\right) - (\psi - \xi) e^{\xi} \mathrm{erfc}\left(\frac{\psi-\xi}{\sqrt{2\psi}}\right)}{e^{\xi} \mathrm{erfc}\left(\frac{\psi+\xi}{\sqrt{2\psi}}\right) + e^{-\xi} \mathrm{erfc}\left(\frac{\psi-\xi}{\sqrt{2\psi}}\right)} \right] \tag{1.68}
$$

where $\xi = v/a$, $\psi = \frac{\sigma_i^2}{\sigma_d^2}$ and the function $\mathrm{erfc}(x) = \left(\frac{2}{\sqrt{\pi}}\right)\int_x^{\infty} e^{-t^2} dt$ is the complementary error function.

Maximum likelihood estimator:

The maximum likelihood of s for given observation v is the value for which the likelihood function $f_{v \mid s}(v \mid s)$ is maximum.

$$
\hat{s} \triangleq \arg\max_s f_{v \mid s}(v \mid s) = \arg\max_s f_{s,v}(s,v) \tag{1.69}
$$

$$
= \arg\min_s \left(\frac{|s|}{a} + \frac{|v-s|^2}{2\sigma^2} \right) \tag{1.70}
$$

$$
= \begin{cases} v - \dfrac{\sigma^2}{a} & \text{if } v \geq \dfrac{\sigma^2}{a} \\[4mm] 0 & \text{if } |v| \leq \dfrac{\sigma^2}{a} \\[4mm] v + \dfrac{\sigma^2}{a} & \text{if } v \leq \dfrac{-\sigma^2}{a} \end{cases} \tag{1.71}
$$

Low Distortion

Let $x(t)$ be the clean speech signal, $n(t)$ be the noise signal, and $y(t)$ be the noisy speech signal.

$$
y(t) = x(t) + n(t) \tag{1.72}
$$

Assume the DCT components of noisy speech, noise, and clean speech signals are represented by $Y(k)$, $N(k)$, and $X(k)$, respectively, where k denotes the position of coefficients in the transform (Soon and Koh 2000). These signals are related in DCT transform as follows:

$$
Y(k) = X(k) + N(k) \tag{1.73}
$$

Let a multiplicative filter be denoted as

$$
\widehat{X}(k) = W(k)Y(k) \tag{1.74}
$$

The expression for $W(k)$ is obtained by minimizing the mean square error D_m which is defined as

$$
D_m = E\left[(W(k)Y(k) - X(k))^2 \right] \tag{1.75}
$$

$$
\approx (W(k) - 2W(k) + 1)E[X(k)^2] + 2\,W(k)(W(k) - 1)E[X(k)N(k)]
$$
$$
+ W(k)^2 E(N(k)^2) \tag{1.76}
$$

Minimizing of D_m will result in the following equation:

$$
W(k) = \frac{\xi(k)}{\xi(k) + 1} \tag{1.77}
$$

where $\xi(k) = \dfrac{E\left[X(k)^2\right]}{E\left[N(k)^2\right]}$

By considering the above filter, the mean square error can be defined as

$$D_m{}^w(k) = \frac{E\left[X(k)^2\right]}{\xi(k) + 1} \tag{1.78}$$

The gain of the Weiner filter is less than one. Sometimes, if $X(k)$ and $N(k)$ are of different sizes, it is possible that $Y(k)$ can be less than $X(k)$. In such case, the attenuation only leads to further distortion of speech. But, the attenuation filter is appropriate when $X(k)$ and $N(k)$ are of same signs. It is assumed that a sign detector can be used to estimate whether $X(k)$ and $N(k)$ are of same sign. Otherwise, two different filters could be derived for these two different cases.

If $X(k)N(k) > 0$

$$E[X(k)N(k)] = E[|X(k)|][|N(k)|] \tag{1.79}$$

$$= \frac{2}{\pi}\sigma_X(k)\sigma_N(k) \tag{1.80}$$

If $X(k)N(k) < 0$

$$E[X(k)N(k)] = -E[|X(k)|][|N(k)|] \tag{1.81}$$

$$= -\frac{2}{\pi}\sigma_X(k)\sigma_N(k) \tag{1.82}$$

Minimizing $D(k)$ using the above equations, we get
If $X(k)N(k) > 0$

$$W(k) = \frac{\xi(k) + \frac{2}{\pi}\sqrt{\xi(k)}}{\xi(k) + \frac{4}{\pi}\sqrt{\xi(k)} + 1} \tag{1.83}$$

If $X(k)N(k) < 0$

$$W(k) = \frac{\xi(k) - \frac{2}{\pi}\sqrt{\xi(k)}}{\xi(k) - \frac{4}{\pi}\sqrt{\xi(k)} + 1} \tag{1.84}$$

Using this set of filters, the mean square error $D_m{}^N(k)$ can be written as

$$D_m{}^N(k) = \frac{(\pi^2 - 4)(\xi(k) + 1)E\left[X(k)^2\right]}{\pi^2\xi(k)^2 + (2\pi^2 - 16)\xi(k) + \pi^2} \tag{1.85}$$

The percentage improvement in the mean square error or the new filter versus the Wiener filter is given as

$$\delta D = \frac{D_m{}^W(k) - D_m{}^N(k)}{D_m{}^W(k)} \tag{1.86}$$

Subtractive Filter

The subtractive filter used by Boll et al. (1979) is another popular method of filtering in the transform domain. The approach used earlier can be repeated here. For subtractive filtering:

$$\widehat{X}(k) = Y(k) - T(k) \tag{1.87}$$

where $T(k)$ is the threshold. The subtractive least mean square error, $D_s(k)$ is defined as

$$D_s(k) = E\left[N(k)^2\right] - 2T(k)E\left[N(k)^2\right] + T(k)^2 \tag{1.88}$$

Minimizing $D_s(k)$ with respect $T(k)$ will result in the following:

$$T(k) = E[N(k)] \tag{1.89}$$

Since the noise variance is assumed as zero, the above equation is not meaningful. However if a sign estimator is available to estimate the sign of $N(k)$, then more meaningful result can be obtained as shown below:

If $N(k) \geq 0$

$$T(k) = E[|N(k)|] = \sqrt{\frac{2}{\pi}}\sigma_N(k) \tag{1.90}$$

If $N(k) < 0$

$$T(k) = -E[|N(k)|] = -\sqrt{\frac{2}{\pi}}\sigma_N(k) \tag{1.91}$$

Using the above filter, the mean square error is given by

$$D_s{}^N(k) = \left(1 - \frac{2}{\pi}\right)\sigma_N^2 \tag{1.92}$$

This could be compared with the conventional subtractive filter implementation given as follows:

$$\widehat{X}(k) = \text{sign}\,(Y(k))(|Y(k)| - T(k)) \tag{1.93}$$

The mean square error for the above equation can be given as:

$$D_s^C(k) = \left(1 + \frac{2}{\pi}\right)\sigma_N^2 \tag{1.94}$$

Low Distortion Nonlinear

The low distortion speech enhancement method is able to identify whether the additive noise in particular frequency is constructive or destructive (Mahmmod et al. 2017). Based on this method multiplicative and subtractive filters has been derived. This approach is useful to reduce the musical noise in speech absence. Assuming the Gaussian statistical model for the clean and noisy signal DCT coefficients, the instances of random processes X_k and Y_k are denoted as x_k and y_k, respectively. Then by using their Joint probability function, we get

$$p_{XY}(x_k, y_k) = \frac{1}{2\pi\sigma_d\sigma_x} \exp\left[-\frac{x_k^2}{2\sigma_x^2} - \frac{(y_k - x_k)^2}{2\sigma_d^2}\right] \tag{1.95}$$

where σ_x^2 and σ_d^2 are the signal and noise variances in the kth DCT coefficient, respectively.

The MMSE estimator of X_k is given as

$$\widehat{X}(k) = E[X_k|Y_k] \tag{1.96}$$

If $\widehat{X}(k)$ is reduced to normal Weiner filter, then this case is for linear estimator. If we assume a polarity detector is available for

$$\widehat{X}_k^{DMMSE} = p_k E\{X_k|Y_k + H_+\} + (1 - p_k)E\{X_k|Y_k + H_-\} \tag{1.97}$$

where p_k is defined as $p_k = \begin{cases} 1 & \text{if event } H_+ \text{ is detected} \\ 0 & \text{if event } H_- \text{ is detected} \end{cases}$

The dual MMSE gain functions for the required filter are derived as follows:

Since Y_k is constructed from the two mutually exclusive events H_+ and H_-, we can write

$$E\{X_k|Y_k + H_+\} = \int_{-\infty}^{\infty} x_k p(x_k|Y_k + H_+)dx_k \tag{1.98}$$

$$= \int_{-\infty}^{\infty} x_k \frac{p(x_k, Y_k, H_+)}{p(Y_k, H_+)} dx_k \tag{1.99}$$

$$E\{X_k|Y_k + H_-\} = \int_{-\infty}^{\infty} x_k \frac{p(x_k, Y_k, H_-)}{p(Y_k, H_-)} dx_k \tag{1.100}$$

The quantities in the above equations can be determined when we know the joint probability functions $p(x_k, Y_k, H_+)$, $p(x_k, Y_k, H_-)$, $p(Y_k, H_+)$, and $p(Y_k, H_-)$. For event

H_+, $X_k D_k > 0$, which results in $X_k Y_k > X_k^2$. By simplifying this condition we get $m_k Y_k > |X_k|$, where $m_k = $ sgn (X_k). Here, sgn$(\cdot)$ is given as

$$
\text{sgn}(\tau) = \begin{cases} +1 & \text{for } \tau > 0 \\ -1 & \text{for } \tau > 0 \\ 0 & \text{for } \tau = 0 \end{cases} \tag{1.101}
$$

The event H_- results in the condition $m_k Y_k < |X_k|$. Thus the joint density functions can be written as

$$
p(x_k, y_k, H_+) = \begin{cases} p_{XY}(x_k, y_k), & m_k Y_k > |X_k| \\ 0 & \text{otherwise} \end{cases} \tag{1.102}
$$

$$
p(x_k, y_k, H_-) = \begin{cases} p_{XY}(x_k, y_k), & m_k Y_k < |X_k| \\ 0 & \text{otherwise} \end{cases} \tag{1.103}
$$

The probability densities $p(Y_k, H_+)$ and $p(Y_k, H_-)$ are defined as

$$
p(Y_k, H_+) = \int_{-\infty}^{\infty} p_{XY}(x_k, y_k, H_+) dx_k \tag{1.104}
$$

$$
p(Y_k, H_+) = \begin{cases} \int_0^{Y_k} p_{XY}(x_k, Y_k) dx_k & \text{if } Y_k \geq 0 \\ \int_{Y_k}^0 p_{XY}(x_k, Y_k) dx_k & \text{if } Y_k < 0 \end{cases} \tag{1.105}
$$

$$
= \text{sgn}(Y_k) \int_0^{Y_k} p_{XY}(x_k, Y_k) dx_k \tag{1.106}
$$

Substituting the values of $p_{XY}(x_k, Y_k)$ Eq. (1.95) into Eq. (1.106), we get

$$
p(Y_k, H_+) = \text{sgn}(Y_k) \frac{e^{-Y_k^2/2(\sigma_d^2 + \sigma_x^2)} [\text{erf}(f_1) + \text{erf}(f_2)]}{\sqrt{2\pi}\,\sqrt{\sigma_d^2 + \sigma_x^2}} \tag{1.107}
$$

where

$$
f_1 = \frac{Y_k \sigma_d}{\sqrt{2}\sigma_x \sqrt{\sigma_d^2 + \sigma_x^2}} \tag{1.108}
$$

$$f_2 = \frac{Y_k \sigma_d}{\sqrt{2}\sigma_d \sqrt{\sigma_d^2 + \sigma_x^2}}$$ (1.109)

and erf($\cdot$) is the error function denoted as

$$\mathrm{erf}(x) = \frac{2}{\sqrt{\pi}} \int_0^x e^{-t^2} dt$$ (1.110)

In the same manner, $p(Y_k, H_-)$ can be shown as

$$p(Y_k, H_-) = \mathrm{sgn}\,(Y_k) \frac{e^{-Y_k^2/2\left(\sigma_d^2 + \sigma_x^2\right)}[\mathrm{erfc}(f_1) + \mathrm{erfc}(f_2)]}{2\sqrt{2\pi}\,\sqrt{\sigma_d^2 + \sigma_x^2}}$$ (1.111)

where erfc($\cdot$) is the complimentary error function given by

$$\mathrm{erfc}(x) = \frac{2}{\sqrt{\pi}} \int_x^\infty e^{-t^2} dt = 1 - \mathrm{erf}(x)$$ (1.112)

1.4   Organization of the Book

The primary aim of the book is to introduce the reader to the use of fractional Fourier transform techniques to enhance the speech quality in both the single and dual channel speech enhancement systems.

This book consists of six chapters. Chapter 1 is introduction to speech enhancement algorithms. Elaborate discussion has been made with regard to time domain and transform domain enhancement techniques. Various algorithms have been presented in detail which are based on time, DFT, DWT, KLT, and DCT domains.

In Chap. 2, an overview of fractional Fourier transform is provided. Theory of FrFT and its properties have been discussed. The discrete fractional Fourier transforms (DFrFT) is reviewed. Various methods for the implementation of DFrFT algorithms are also discussed. This chapter also discusses the limitations and advantages of the each method of implementation for DFrFT.

In Chap. 3, a new approach for speech enhancement is presented based on the adaptive noise cancellation using FrFT. This chapter explores how the FrFT algorithm is employed in adaptive noise cancellation (ANC) for speech enhancement and its improved performance over the time and frequency domain adaptive filtering techniques.

In Chap. 4, the discrete fractional cosine transform (DFrCT) is introduced to single channel speech enhancement system. A Weiner filter with harmonic regeneration noise reduction techniques has been implemented in DFrCT domain. The proposed algorithm is studied for real-world noise conditions at various input SNR levels and the results are compared with conventional Weiner filtering approach for enhancement. The performance of algorithm is analyzed both in terms of quality and intelligibility of enhanced speech.

In Chap. 5, a new speech enhancement approach based on discrete fractional sine transform (DFrST) algorithm is presented. Results are compared with standard Wiener filter and Wiener filter with DFrCT techniques.

Chapter 6 summarizes the work presented in this book, highlights the main contributions of the work, draws the conclusions, and provides suggestions for future work. Finally, the references used in this research work are listed.

1.5 Conclusions

In this chapter, most commonly used speech enhancement algorithms in time domain and transform domain are discussed. As discussed in Sect. 1.1, fractional Fourier transform is also the possible transform domain in which speech signal can be enhanced through the spectral modification. So far, very few studies have reported fractional Fourier transform techniques for enhancing speech. In this book, fractional Fourier transform techniques are employed for speech enhancement. A detail explanation of the FrFT-based speech enhancement algorithms is provided in the next chapter.

References

Boll, S. F. (1979). Suppression of acoustic noise in speech using spectral subtraction. *IEEE Transactions on Acoustics, Speech, and Signal Processing, 27*(2), 113–120. https://doi.org/ 10.1109/TASSP.1979.1163209

Ephraim, Y., & Malah, D. (1984). Speech enhancement using a minimum-mean square error short-time spectral amplitude estimator. *IEEE Transactions on Acoustics, Speech, and Signal Processing, 32*(6), 1109–1121. https://doi.org/10.1109/TASSP.1984.1164453

Ephraim, Y., & Malah, D. (1985). Speech enhancement using a minimum mean-square error log-spectral amplitude estimator. *IEEE Transactions on Acoustics, Speech, and Signal Processing, 33*(2), 443–445. https://doi.org/10.1109/TASSP.1985.1164550

Ephraim, Y., & Van Trees, H. L. (1995). A signal subspace approach for speech enhancement. *IEEE Transactions on Speech and Audio Processing, 3*(4), 251–266. https://doi.org/10.1109/89. 397090

Lim, J., Oppenheim, A., & Braida, L. (1978). Evaluation of an adaptive comb filtering method for enhancing speech degraded by white noise addition. *IEEE Transactions on Acoustics, Speech, and Signal Processing, 26*(4), 354–358. https://doi.org/10.1109/TASSP.1978.1163117

Mahmmod, B. M., Ramli, A. R., Abdulhussian, S. H., Al-Haddad, S. A. R., & Jassim, W. A. (2017). Low-distortion MMSE speech enhancement estimator based on Laplacian prior. *IEEE Access, 5*, 9866–9881. https://doi.org/10.1109/ACCESS.2017.2699782

Ram, R., & Mohanty, M. N. (2018). The use of deep learning in speech enhancement. In *Proceedings of the First International Conference on Information Technology and Knowledge Management* (Vol. 14, pp. 107–111). https://doi.org/10.15439/2017km40

Soon, I. Y., & Koh, S. N. (2000). Low distortion speech enhancement. *IEE Proceedings: Vision, Image and Signal Processing, 147*(3), 247–253. https://doi.org/10.1049/ip-vis:20000323

Soon, I. Y., Koh, S. N., & Yeo, C. K. (1998). Noisy speech enhancement using discrete cosine transform. *Speech Communication, 24*(3), 249–257. https://doi.org/10.1016/S0167-6393(98)00019-3

Wang, Y., & Wang, D. (2015). A deep neural network for time-domain signal reconstruction. In *ICASSP, IEEE international conference on acoustics, speech and signal processing - proceedings, 2015-August* (pp. 4390–4394). https://doi.org/10.1109/ICASSP.2015.7178800

Yu, H., Ouyang, Z., Zhu, W. P., Champagne, B., & Ji, Y. (2019). A deep neural network based Kalman filter for time domain speech enhancement. In *Proceedings - IEEE International Symposium on Circuits and Systems, 2019-May*. https://doi.org/10.1109/ISCAS.2019.8702161

Zhang, W., Benesty, J., & Chen, J. (2016). Single-channel noise reduction via semi-orthogonal transformations and reduced-rank filtering. *Speech Communication, 78*, 73–83. https://doi.org/10.1016/j.specom.2015.12.007

Zou, X., & Zhang, X. (2007). Speech enhancement using an MMSE short time DCT coefficients estimator with supergaussian speech modeling. *Journal of Electronics, 24*(3), 332–337. https://doi.org/10.1007/s11767-005-0174-y

Chapter 2
Fractional Fourier Transform

2.1 Theory of Fractional Fourier Transform

The fractional Fourier transform is the generalization of the conventional Fourier transform (FT) and can be interpreted as a counterclockwise rotation of the signal to any angles in the time-frequency plane (Almeida 1994; Cariolaro et al. 1998; Mendlovic and Ozaktas 1993; Ozaktas and Mendlovic 1993). The p-th order continuous FrFT of a signal $x(t)$ is defined as:

$$F_p[x(t)](u) \equiv X_p(u) = \int_{-\infty}^{\infty} x(t) B_p(t, u) dt \qquad (2.1)$$

where $B_p(t, u)$ is the continuous FrFT Kernel given by

$$B_p(t, u) = K_\varphi \exp\left[j\{(t^2 + u^2)/2\} \cot \alpha - jut \operatorname{cosec} \alpha \right] \qquad (2.2)$$

and $K_\varphi = \sqrt{(1 - j \cot \alpha)/2\pi}$

The transformation angle $\alpha = p\pi/2$ for $0 < |p| < 2$.

From the point of view of signal analysis, the major property of the FrFT is the property of rotation in time-frequency plane (TFp). FrFT rotates the signal to be transformed in a counterclockwise direction on the time-frequency plane by an angle which is proportional to the transformation order p. The geometrical illustration of FrFT is shown in Fig. 2.1.

From the above figure, t_p and f_p are always orthogonal and $t_{p+2} = f_{p+1} = -t_p$, and $t_{p-1} = -f_p$.

© The Author(s), under exclusive licence to Springer Nature Switzerland AG 2020   25
P. Kunche, N. Manikanthababu, *Fractional Fourier Transform Techniques for Speech Enhancement*, SpringerBriefs in Speech Technology,
https://doi.org/10.1007/978-3-030-42746-7_2

Fig. 2.1 Illustration of time-frequency representation of FrFT

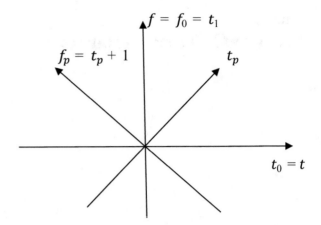

2.1.1   Properties of FrFT

1. Reduces to Classical Fourier Transform: when the transform order $p = 1$, the FrFT reduces to the normal FT.
2. Parity Operator: when α is a multiple of π, FrFT degenerates into parity and identity operators.
3. Identity Operator: It is also called zero rotation. If $p = 0$, $F^p = I$, where I is the identity operator.
4. Index Additive: FrFT is additive in index ($F^p F^q = F^{p+q}$).
5. Square root of F: It is a special case of FrFT, we have $F^{\frac{1}{2}} F^{\frac{1}{2}} = F$ and $F^{\frac{1}{2}}$ is known as the square root of F.
6. Linear: FrFT is a linear operator.
7. Unitary: $(F^p)^{-1} = F^{-p} = (F^p)^{\dagger}$, where $(.)^{\dagger}$ is the Hermitian conjugation.
8. Associative: $(F^r F^q) F^p = F^r (F^q F^p)$
9. Commutative: $F^p F^q = F^q F^p$.
10. Conservation of symmetry: The FrFT of an even function is even and odd function is odd.
11. Parseval: Parseval's relation is preserved between the time or spatial domain and the fractional domain.

$$\int_{-\infty}^{\infty} f(u)\overline{g(u)}du = \int_{-\infty}^{\infty} (F^p f)(t)\overline{(F^p g)(t)}dt$$

If $f = g$, it is called energy conservation property.
12. Time Reversal: $F^p P = P F^p$

The additional properties of FrFT and FrFT for some standard signals are given in Tables 2.1 and 2.2, respectively. The FrFTs of different rotation angles of a cosine signal is illustrated in Fig. 2.2. Figure. 2.2a is the FrFT when the order of transform $p = 0.05$, Fig. 2.2b is the FrFT when order $p = 0.25$, Fig. 2.2c is the FrFT when order

Table 2.1 Additional properties of fractional Fourier transform

Aperiodic signal	FrFT with angle α	Description
$ax(t) + by(t)$	$aX_\alpha(u) + bY_\alpha(u)$	Linearity
$x(t - T)$	$e^{j\frac{T^2}{2}\cot\alpha}e^{-juT\csc\alpha}X_\alpha(u - T\cos\alpha)$	Time shift
$e^{j2\pi Ft}x(t)$	$e^{-j\frac{v^2\sin\alpha\cos\alpha}{2}+juv\cos\alpha}X\alpha(u - v\sin\alpha)$	Modulation
$\frac{d}{dt}x(t)$	$\cos\alpha\frac{d}{du}X\alpha(u) + ju\sin\alpha X_\alpha(u)$	Derivative
$\int_{-\infty}^{t}x(t)dt$	$\sec\alpha\cdot e^{-\frac{j}{2}u^2\tan\alpha}\int_{-\infty}^{u}e^{\frac{j}{2}v^2\tan\alpha}X\alpha(v)dv$	Integration
$tx(t)$	$u\cos\alpha X\alpha(u) + j\sin\alpha\frac{d}{du}X\alpha(u)$	Time multiplication

Table 2.2 FrFT of related and some common signal

FrFT with angle α	Condition
$\delta(t - \tau)$	$\sqrt{\frac{1-j\cot\alpha}{2\pi}}\ e^{j\left((\tau^2+u^2)/2\right)\cot\alpha - ju\tau\csc\alpha}$
1	$\sqrt{1+j\tan\alpha}\ e^{-j(u^2/2)\tan\alpha}$
e^{jvt}	$\sqrt{1+j\tan\alpha}\,e^{-j\left((v^2+u^2)/2\right)\tan\alpha+juv\sec\alpha}$
$e^{jc(t^2/2)}$	$\sqrt{\frac{1+j\tan\alpha}{1+c\tan\alpha}}\ e^{-j(u^2/2)(c-\tan\alpha)/(1+c\tan\alpha)}$
$e^{-(t^2/2)}$	$e^{-(u^2/2)}$
$H_n e^{-(t^2/2)}$ H_n–Hermite polynomial	$e^{-jn\alpha}H_n e^{-(u^2/2)}$
$e^{-c(t^2/2)}$	$\sqrt{\frac{1-j\cot\alpha}{c-j\cot\alpha}}\ e^{-j(u^2/2)(c^2-1)\cot\alpha/(c^2+\cot^2\alpha)}\times$ $e^{-j(u^2/2)\ (c^2-1)\cot\alpha/(c^2+\cot^2\alpha)}$
$\cos(vt)$	$\sqrt{1+j\tan\alpha}\cdot e^{-\frac{j}{2}(u^2+v^2)\tan\alpha}\cos(uv\sec\alpha)$
$\sin(vt)$	$\sqrt{1+j\tan\alpha}\cdot e^{-\frac{j}{2}(u^2+v^2)\tan\alpha}\sin(uv\sec\alpha)$
$e^{-j(at^2+bt+c)}$	$\sqrt{\frac{1-j\cot\alpha}{j\cdot 2a-j\cot\alpha}}e^{j\frac{2a\cot\alpha-1}{\cot\alpha-2a}u^2}e^{-j\frac{b\csc\alpha}{\cot\alpha-2a}e}-j\frac{b^2}{2(\cot\alpha-2a)}-jc$

$p = 0.5$, Fig. 2.2d is the FrFT when order $p = 0.75$, and Fig. 2.2e is the FrFT when order $p = 1$.

2.2 Discrete Fractional Fourier Transform

In order to practically compute the FrFT, due to its fast oscillations of the quadratic complex exponential kernels, it is not possible to implement through direct numerical integrations. Hence several approaches have been proposed so far for the practical computation of FrFT by decomposing these integral transformations into simple sub-operations. The different methods used for DFrFT computation are classified as: Sampling of FrFT, Improved sampling of FrFT, DFrFT through Linear combination, Eigen vector decomposition type, Group theory type and Impulse train

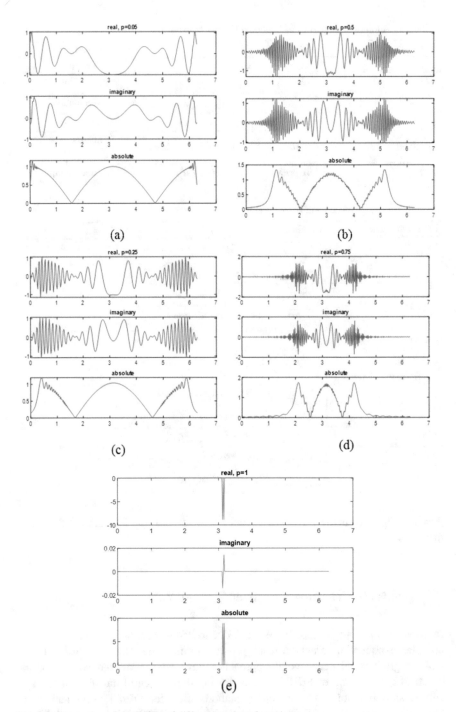

Fig. 2.2 Illustration of the FrFTs of different angles of rotation

type DFrFTs. In this chapter, the background of three main techniques for the computation of DFrFT is provided as follows:

2.2.1 DFrFT Based on Sampling of FrFT

Sampling of FrFT is a simple and straightforward approach to get DFrFT, since the sampling theorems of FrFT follows the Shannon sampling theorem (Candan and Ozaktas 2003; Erseghe et al. 1999; Sharma and Joshi 2005; Tao et al. 2008a; Xia 1996; Zayed 1996; Zayed and García 1999). However, the DFrFT obtained through direct sampling of FrFT cannot hold the closed form properties and it is also not additive and hence the applications are limited.

Ozaktas et al. (1996) proposed DFrFT implementation in two methods. Both of them are derived by manipulating the FrFT expression to get appropriate sampling.

Approach 1 Approach 1 is comprised of three main steps as follows:

1. Multiplication of signal $x(t)$ with a chirp function:

$$g(t) = \exp\left[-j\pi t^2 \tan\left(p/2\right)\right]x(t) \tag{2.3}$$

2. Chirp convolution

$$h(t) = A_\alpha \int_{-\infty}^{\infty} \exp\left[j\pi csc(p)\right] \times (t - \tau)g(\tau)d\tau \tag{2.4}$$

3. Chirp post multiplication

$$F_p(x(t)) = \exp\left[-j\pi t^2 \tan\left(p/2\right)\right]h(t) \tag{2.5}$$

To perform the convolution operation in the above equation, first sample the signal $g(t)$, and apply normal fast Fourier transform (FFT). As the bandwidth and time-bandwidth product of $g(t)$ is twice of that of $x(t)$, the signal $x(t)$ is interpolated by a factor 2.

The above procedure can be given as

$$X_p = F_p x = D\Lambda H\Lambda Jx \tag{2.6}$$

where x and X_p are the column vectors consisting the samples of $x(t)$ and its DFrFT, D and J are the matrices representing the operations of decimation and interpolation, Λ is a diagonal matrix corresponds to multiplication and H corresponds to the convolution operation.

Approach 2 In the second method implemented by Ozekatas et al. (1996), the FrFT expression is rewritten into the form given below:

$$X_p = A_p \exp\left(j\pi \cot(p)u^2\right) \times \int_{-\infty}^{\infty} \exp\left(-j2\pi \csc(p)ut\right)$$

$$\times [\exp\left(j\pi \cot(p)t\right)x(t)]dt \tag{2.7}$$

The Shannon's interpolation formula for the modulated function $[\exp(j\pi \cot(p)t)\ x(t)]$ is given as

$$[\exp\left(j\pi \cot(p)t\right)x(t)] = \sum_{n=-N}^{N}\left[\exp\left(j\pi \cot(p)\frac{n}{2\Delta t}\right)\right.$$

$$\left.\times x\left(\frac{n}{2\Delta t}\right) \sin c\left(2\Delta t\left(t - \frac{n}{2\Delta t}\right)\right)\right] \tag{2.8}$$

Therefore the samples of DFrFT can be obtained in terms of the samples of the original signal as

$$X_p(m) = \frac{A_p}{2\Delta t}$$

$$\times \sum_{n=-N}^{N} \exp\left(j\pi\left(\cot(p)\times\left(\frac{m}{2\Delta t}\right)^2 - 2\csc(p)\frac{mn}{(2\Delta t)^2} + \cot(p)\left(\frac{n}{2\Delta t}\right)^2\right)\right)x\left(\frac{n}{2\Delta t}\right) \tag{2.9}$$

here, Δt is the sampling interval.

The overall process of "Approach 2" is given as

$$X_p = F_p x = DKJx \tag{2.10}$$

where $K(m,n) = \frac{A_a}{2\Delta t}\exp\left(j\pi\left(\cot(\alpha)\left(\frac{m}{2\Delta t}\right)^2 - 2\csc(\alpha)\frac{mn}{(2\Delta t)^2} + \cot(\alpha)\left(\frac{m}{2\Delta t}\right)^2\right)\right)$, for $|m|$ and $|n| \leq N$ and A_α is a constant.

Pei and Ding (Pei and Ding 2000) proposed another type of DFrFT called "Improved sampling type DFrFT." This DFrFT satisfies the properties of unitary and reversible. Furthermore, the authors also gave an analytical expression for the closed form of FrFT.

$$y(n) = x(n\Delta t) \tag{2.11}$$

$$Y_p(m) = X_p(m\Delta u) \tag{2.12}$$

$$Y_p(m) = \sqrt{\frac{\sin p - j\cos p}{2M + 1}} \times \exp\left(\frac{j}{2}\cot p m^2 \Delta u^2\right)$$

$$\times \sum_{n=-N}^{N} \exp\left(-j\frac{2\pi nm}{2M + 1}\right) \times \exp\left(\frac{j}{2}\cot p n^2 \Delta t^2\right) y(n) \qquad (2.13)$$

$$Y_p(m) = \sqrt{\frac{\sin p - j\cos p}{2M + 1}} \times \exp\left(\frac{j}{2}\cot p m^2 \Delta u^2\right)$$

$$\times \sum_{n=-N}^{N} \exp\left(-j\frac{2\pi nm}{2M + 1}\right) \times \exp\left(\frac{j}{2}\cot p n^2 \Delta t^2\right) y(n) \qquad (2.14)$$

2.2.2 Linear Combination Type-DFrFT

Dickinson and Steiglitz (1982) derived DFrFT based on the linear combination of identity operation (F_0), DFT ($F_{\pi/2}$), time inverse operation (F_π), and inverse DFT ($F_{3\pi/2}$). The fractional matrix operator F_p for $0 \le p \le \frac{\pi}{2}$ is formulated as

$$F_p = \sum_{k=0}^{3} \beta_k F_{k\pi/2} \qquad (2.15)$$

where

$$\beta_k = \frac{1}{4}\sum_{l=1}^{4}\exp\left[j\pi l\left(\frac{2p}{\pi}-k\right)/2\right] \qquad (2.16)$$

for $0 \le k \le 3$.
   Properties:

1. Unitary: $F_p F_{-p} = I$
2. Angle additive: $F_p F_q = F_{p+q}$
3. Angle multiplicative: $F_p{}^m = F_{mp}$

   The limitation of linear combination type DFrFT is it is not a discrete version of continuous FrFT (Pei et al. 1998).

2.2.3 Group Theory Type-DFrFT

Group theory DFrFT is derived from the multiplication of DFT with periodic chirps. This DFrFT obeys rotational property of Wigner Distribution (WD), additivity, and reversibility. This algorithm provides relation between the DFT and discrete WD based on group theory.

To rotate a continuous time-frequency distribution by an angle θ, a matrix $R(\theta)$ is to be applied to the coordinates WD and is given as

$$R(\theta) = \begin{bmatrix} \cos\theta & \sin\theta \\ -\sin\theta & \cos\theta \end{bmatrix} \tag{2.17}$$

In case $\theta = \frac{-\pi}{2}$, in time domain it is equal to application of FrFT with an appropriate order $\theta = \frac{-\pi}{2}$.

The application of symplectic transformation A, to a WD, is "$W_{f(x)g(x)}$" which is equivalent to the application of a unitary transformation dependent on A, $(U(A))$, to both $f(x)$ and $g(x)$ first, and then computing their WD (O'Neill and Williams 1996), i.e.,

$$W_{f(x)g(y)}[A(t,f)] = W_{U(A)[f(x)]U(A)[g(x)]}(t,f) \tag{2.18}$$

The symplectic transformation A^{γ} corresponds to counter clockwise rotation by 90° and is given by

$$A^{\gamma} = \begin{bmatrix} 0 & 1 \\ -1 & 0 \end{bmatrix} \tag{2.19}$$

$$U(A^{\gamma})[f(n)] = j^{1/2} \sum_{k=0}^{N-1} e^{\frac{j2\pi nk}{N}} F^{1}[f(n)] \tag{2.20}$$

$$= j^{1/2} \sum_{k=0}^{N-1} e^{\frac{j2\pi nk}{N}} F(k) \tag{2.21}$$

where $F(k)$ is the DFT of $f(n)$, $U(A^{\gamma})$ is proportional to inverse DFT. It is clear that the application of symplectic transformation A^{γ} does not result in a rotation by 90°.

In a continuous time-frequency plane, a rotation by θ is a matrix shown in Eq. (2.17). The set of such rotations in two dimensions comprise an algebraic group called the special orthogonal group of dimension-2 over the real numbers or $SO(2, R)$. The elements $R(\theta)$ of $SO(2, R)$ must satisfy the following conditions (Almeida 1994):

1. Zero rotation: $R(0) = I$
2. Consistency with FT: $R\left(\frac{-\pi}{2}\right) = F^{1}$

3. Additivity: $R(\alpha) + R(\beta) = R(\alpha + \beta)$

The rotation matrix R_P for discrete time discrete frequency of a signal with length P, $SO(2, Z/P)$ is a special orthogonal group of dimension-2 over the integers modulo P. The R_P is given as

$$R_P(a,b) = \begin{bmatrix} a & b \\ -b & a \end{bmatrix} \tag{2.22}$$

where a and b are the elements of integers mod $(P, Z/P)$ and the determinant of $R_P(a, b)$ is equal to1.

2.2.4   Impulse Train Type-DFrFT

Impulse train DFrFT is a special case of continuous FrFT where the input function is considered as an equally spaced periodic impulse train. If in a period Δ_0, the number of impulses is N, then N should be equal to Δ_0^2.

Let $f(x)$ is a sampled periodic signal with a period Δ_0 which is given as (Candan et al. 2000)

$$f(x) = \sum_{k=-N/2}^{N/2-1} f\left(k\frac{\Delta_0}{N}\delta\left[x - \left(n + \frac{k}{N}\right)\right]\Delta_0\right) \tag{2.23}$$

where N is a number of samples in a period taken to be even.

The FrFT of order p can be defined as

$$F_p[f(x)] = \int_{-\infty}^{\infty} K_p(x, x')f(x)dx \tag{2.24}$$

$$= \sum_{k=-N/2}^{N/2-1} f\left(k\frac{\Delta_0}{N}\sum_{n=-\infty}^{\infty} K_p\left[x, \left(n + \frac{k}{N}\right)\right]\Delta_0\right) \tag{2.25}$$

For the transformed function in Eq. (2.25) to be periodic with a period Δ_p, $F_p[f(x)]$ should be equal to $F_p[f(x + l\Delta_p)]$ for all x and l.

The following are the conditions to satisfy the periodicity (Ozaktas et al. 1996):

$$\Delta_p \cos \varnothing = s\Delta_0 \tag{2.26}$$

$$\Delta_p \sin \varnothing = t\frac{N}{\Delta_0} \tag{2.27}$$

where s and t are integers such that stN is even, since N is even.

The condition for transformation order is given as

$$\tan \varnothing = \frac{s}{t}\frac{N}{\Delta_0^2} \tag{2.28}$$

which means $F_p[f(x)]$ is periodic for only a set of orders which satisfies Eq. (2.28)
By using shift property of FrFT,

$$F_p[f(x)] = \sum_{k=-N/2}^{N_2-1} f\left(k\frac{\Delta_0}{N}\right)\exp(B)\sum_{n=-\infty}^{\infty} K_p\left[x - \frac{k}{N}\right]\Delta_0 \tag{2.29}$$

where $B = j\pi\left[\sin\varnothing\cos\varnothing\left(k\frac{\Delta_0}{N}\right)^2\right] - 2\sin\varnothing xk\frac{\Delta_0}{N}$

The inner summation is the shifted FrFT of a uniform impulse train with period Δ_0 and is given by

$$\overline{\delta}_p[f(x)] = \int_{-\infty}^{\infty} K_p(x, x') \sum_{n=-\infty}^{\infty} \delta(x - n\Delta_0) \tag{2.30}$$

$$\overline{\delta}_p[f(x)] = A_\varnothing\sqrt{\frac{2r^2}{stN}} \sum_{n=-\infty}^{\infty} \exp\left(j\pi\cot\varnothing x^2\right)\delta\left(x - \frac{nr}{\Delta_p}\right) \tag{2.31}$$

where r is greatest common divider of s and t.
    If s and t are prime, then

$$F_p[f(x)] = \sqrt{\frac{2r^2}{stN}}A_\varnothing \sum_{k=-N_2}^{N_2-1} f\left(k\frac{\Delta_0}{N}\right)\exp(z)\sum_{n=-\infty}^{\infty}\delta\left(x - \frac{n}{\Delta_p}\right) \tag{2.32}$$

where $z = j\pi\left[x^2\cot\varnothing - 2x\frac{k\Delta_0}{N}\csc\varnothing + \left(k\frac{\Delta_0}{N}\right)^2\cot\varnothing\right]$
    The definition of DFrFT is given as

$$F_p\left[f\left(\frac{n}{\Delta_p}\right)\right] = \sum_{k=-N_2}^{N_2-1} T_\varnothing\left[\frac{n}{\Delta_p}, \frac{k}{\Delta_0}\right]f\left(\frac{k}{\Delta_0}\right) \tag{2.33}$$

where $N = \Delta_0^2$ is used to get the symmetric form of the transform and the discrete transformation kernel $T_\varnothing$ is given as

$$T_\varnothing\left(\frac{n}{\Delta_p}, \frac{k}{\Delta_0}\right) = \sqrt{\frac{2}{stN}}A_\varnothing\exp\left(j\pi(z)\right) \tag{2.34}$$

where $z = \left(\frac{n}{\Delta_p}\right)^2 \cos\varnothing - 2\frac{n}{\Delta_p}\frac{k}{\Delta_0}\csc\varnothing + \cot\varnothing + \left(\frac{k}{\Delta_0}\right)^2 \cot\varnothing$

2.2.5 Eigenvector Decomposition Type-DFrFT

Eigenvector decomposition type-DFrFT was proposed by Pei et al. (1999). This DFrFT searches the eigenvectors set of the N point DFT and then compute the fractional power of DFT matrix. The limitations of previously discussed sampling type-DFrFT such as non-unitary and non-additivity can be combat with this type of DFrFT.

In this method, DFrFT is calculated based on eigen decomposition of DFT kernel matrix. The transform kernel of DFrFT is defined as

$$F_{2\pi/\alpha} = U D_{2\pi/\alpha} U \tag{2.35}$$

$$F_{2\pi/\alpha} = \sum_{k=0}^{N-1} \exp(-jk\alpha) \times u_k u_k^T \qquad \text{for } N \text{ is odd} \tag{2.36}$$

$$= \sum_{k=0}^{N-2} \exp(-jk\alpha) \times U_k U_k^T$$

$$+ \exp(-jN\alpha) U_N U_N^T \qquad \text{for } N \text{ is even} \tag{2.37}$$

here $U = [u_0 u_1 \ldots u_{N-1}]$, when N is odd, and $U = [u_0 u_1 \ldots u_{N-2}]$, when N is even.

Is the normalized vector corresponding to the kth order Hermite function and is given as

$$D_{2\pi/\alpha} = \text{diag}\left(\exp(-j0), \exp(-j\alpha) \ldots \exp(-j\alpha(N-2)), \exp(-j\alpha(N-1))\right) \text{ for } N \text{ odd} \tag{2.38}$$

and

$$D_{2\pi/\alpha} = \text{diag}\left(\exp(-j0), \exp(-j\alpha) \ldots \exp(-j\alpha(N-2)), \exp(-j\alpha(N-1))\right) \text{ for } N \text{ even} \tag{2.39}$$

Candan et al. (2000) proposed another approach for computation of DFrFT based on the definitions proposed by Pei and Yeh (1997).

The definition for DFrFT in (Candan et al. 2000) is as follows:

Table 2.3 Eigenvalues multiplicities of DFT kernel matrix

N	1	$-j$	-1	j
$4m$	$m + 1$	m	m	$m - 1$
$4m + 1$	$m + 1$	m	m	m
$4m + 2$	$m + 1$	m	$m + 1$	m
$4m + 3$	$m + 1$	$m + 1$	$m + 1$	m

$$F_p[m, n] = \sum_{k=0, k \neq (N-1+(N)_2)}^{N} u_k[m] e^{-j\frac{\pi}{2}kp} u_k[n] \tag{2.40}$$

where $u_k[n]$ represents the kth discrete Hermite–Gaussian function (the eigenvector of S with k zero crossings) and $(N)_2 \equiv N \bmod 2$.

The range of the summation given in the above equation is because of the fact that there is no eigenvector with $N - 1$ or N zero crossings when N is even or odd, respectively. The eigenvalue multiplicity of DFT kernel matrix is given in Table 2.3.

Algorithm for DFrFT computation is as follows:

1. Generate matrices S and P.
    The S matrix is given as

$$S = \begin{bmatrix} 2 & 1 & 0 & \cdots & 0 & 1 \\ 1 & 2\cos(\omega) & 1 & \cdots & 0 & 0 \\ 0 & 1 & 2\cos(2\omega) & \cdots & 0 & 0 \\ \vdots & \vdots & \vdots & \ddots & \vdots & \vdots \\ 0 & 0 & 0 & \cdots & 2\cos[(N-2)\omega] & 1 \\ 1 & 0 & 0 & \cdots & 1 & 2\cos[(N-1)\omega] \end{bmatrix}$$

$$\tag{2.41}$$

The P matrix is defined below for P odd and even. This matrix decomposes an arbitrary vector $f(n)$ into its even and odd components. The P matrix maps the even part of the N dimensional vector to the first $\left(\frac{N}{2} + 1\right)$ components and the odd part to the remaining components. For example, the P matrix of dimension 5 is given as

$$P = \frac{1}{\sqrt{2}} \begin{bmatrix} 2 & 0 & 0 & 0 & 0 \\ 0 & 1 & 0 & 0 & 1 \\ 0 & 0 & 1 & 1 & 0 \\ 0 & 0 & 1 & -1 & 0 \\ 0 & 1 & 0 & 0 & 1 \end{bmatrix} \tag{2.42}$$

2. Generate E_v and O_d matrices

$$PSP^{-1} = PSP = \begin{pmatrix} E_v & 0 \\ 0 & O_d \end{pmatrix} \tag{2.43}$$

3. Calculate the eigenvectors and eigenvalues of E_v and O_d
4. Sort the eigenvalues of $E_v(O_d)$ in the descending order of eigenvalues of $E_v(O_d)$ and denote the sorted eigenvectors as $e_k(o_k)$
5. The contents of U matrix are calculated by following equations:

$$u_{2k}[n] = P\left[e_k^T|0.....0|\right]^T \tag{2.44}$$

and

$$u_{2k+1}[n] = P\left[|0.....0|o_k^T\right]^T \tag{2.45}$$

6. Define the FrFT operator matrix as

$$F^P(m,n) = \sum_{k \in \mu} u_k[m] e^{-j\frac{\pi k p}{2}} u_k[n] \tag{2.46}$$

$$\mu = \left\{0, \ldots, N-2, \left(N-(N)_2\right)\right\} \tag{2.47}$$

7. The order of the FrFT can be calculated from the below equation

$$f^p = F^p f \tag{2.48}$$

Some of the essential properties of DFrFT are listed as follows (Candan et al. 2000):

1. $f[n] \overset{P}{\leftrightarrow} f_p[n]$
2. $f[n] + g[n] \overset{P}{\leftrightarrow} f_p[n] + g_p[n]$
3. $f_p[n] \overset{q}{\leftrightarrow} f_{p+q}[n]$
4. $f[n] \overset{p=1}{\leftrightarrow} \text{DFT}\{f[n]\}$
5. $f[-n] \overset{P}{\leftrightarrow} f_p[-n]$
6. $f^*[n] \overset{P}{\leftrightarrow} f_p^*[-n]$
7. even $\{f[n]\} \overset{P}{\leftrightarrow}$ even $\{f_p[n]\}$
8. odd $\{f[n]\} \overset{a}{\leftrightarrow}$ odd $\{f_p[n]\}$

Table 2.4 Comparison for different types of DFrFT

Type of DFrFT	Advantages	Limitations
Direct computation	Easy to design No constraints except ad-bc = I	Does not hold many important properties. • Not be unitary • Not be additive • Not be reversable
Improved sampling	Sample the continuous FrFT properly Similar to the continuous case It is a fast algorithm	Kernel is not orthogonal and additive Many constraints Cannot be calculated for all transform orders
Linear combination	Transform kernel is orthogonal Angle additive Time reversible	Very similar to continuous DFT Lose important characteristic of fractionalization
Group theory	Holds the property of additivity Holds the property of reversable	It can be calculated for only some specific angles Number of points N is prime
Impulse train	Many properties of FrFT exists	Many constraints Not defined for all values of p
Eigen vector decomposition	Good in removing chirp noise Less constraints compared to other approaches Smallest deviation factor	High computational complexity

9. $\displaystyle\sum_{n=0}^{N-1} |f[n]|^2 = \sum_{n=0}^{N-1} |f_p[n]|^2$

The advantages and limitations of each of the DFrFT implementation type is given in Table 2.4.

2.3 Advantages of FrFT

- The FrFT is more general and flexible than the FT.
- The FrFT can be applied to partial differential equations (order $n > 2$). If we choice appropriate parameter p, then the equation can be reduced order to $n - 1$.
- The Fourier transform only deal with the stationary signals; we can use the FrFT to deal with time-varying signals.
- Using the FrFT to design the filters, it can reduce the MMSE (minimum mean square error). Besides, using the FrFT, many noises can be filtered out that the FT cannot remove in optical system, microwave system, radar system, and acoustics.
- In encryption applications it is safe to use FrFT than conventional FT, since FrFT has more parameters that FT.

2.4 Applications of FrFT

2.4.1 Optimal Filtering

Let us consider the case shown in Fig. 2.3, here the original signal and the undesired noise terms are presented on the same TFp. As the original signal is distributed on the TFp, naturally its projection overlaps with the noise projection on both time (t) and frequency (f) axes, thus it is very challenging to separate the original signal from the undesired components in time domain when the order of transformation is zero or in frequency domain when the order of transformation is one, respectively. Nevertheless, due to the reason the time/space frequency plane is rotated in FrFT, it can divide the energy overlap between the signal and noise. Especially for transient and non-stationary signal processing, the FrFT can be used to filter signals with heavy noise terms (Ashok Narayanan and Prabhu 2003; Kutay et al. 1997; Erden et al. 1999; Ozaktas et al. 1994).

2.4.2 Image Processing

FrFT has been widely used in the research of image processing. DFrFT has been applied for color image encoding and decoding in combination with discrete wavelet transform (DWT). FrFT has also been proved as a potential time-frequency tool to represent the time-frequency characteristic of an image and the distribution changes with the variety of the fractional powers. FrFT can also be applied in the area of image recognition and image edge extraction. Image compression using FrFT has been reported by (Singh 2006). The use of FrFT has been explored for image encryption. Guo et al. (2011) proposed a watermarking algorithm for optical images by the combination of random phase encoding and FrFT. Li et al. (2012) proposed a method for log-polar coordinates coefficients (LPCC) estimation by the method of

Fig. 2.3 Illustration of signal filtering in the FrFD

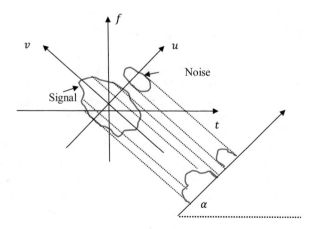

Multilayer-pseudo polar FrFT (MPFFT). Afterwards, Li et al. (2013) proposed a novel method to estimate LPCC using MPFFT, in order to acquire better image registration accuracy.

2.4.3   Signal Analysis

FrFT has been successfully applied for signal sampling which plays a crucial role in digital signal processing. In the conventional sampling processing, the sampled signal is always analyzed in the Fourier transform (FT) domain. In Jin et al. (2016), authors applied the fractional Fourier transform (FrFT) analysis into sampling processing. In contrast to the FT-based analysis method, the new sampling method developed with sub-Nyquist rate based on FrFT achieved high SNR and low sampling rate of an analog-to-digital converter (ADC).

2.4.4   Pattern Recognition

FrFT is used for target classification from the synthesized active sonar returns from targets (Seok and Bae 2014). It is used to extract the shape variation in FrFD based on the aspects of the target. Neural network is used to classify the extracted features in FrFD (Barshan and Ayrulu 2002).

The fractional Fourier transform (FrFT) time-frequency frame work for flaw identification and classification is developed using an inhomogeneous wave equation where the forcing function is prescribed as a linear chirp, modulated by a Gaussian envelope (Tant et al. 2015). It is noted that the FrFT does not change the scattering profile of flaw, but it is amplified. Hence the FrFT-based method enhanced the signal to noise ratio (SNR).

2.4.5   Optical Engineering

FrFT has been introduced in the optical engineering. The flat-topped multi-Gaussian beam (FMGB) FrFT has been investigated by Gao et al. (2010). They studied three types of Fourier transform (FrFT) based optical systems (OSs) such as Lohmann I & II, and quadratic graded-index systems (Han et al. 2012). Zhao and Cai (2010) studied general-type beam paraxial propagation through a truncated FrFT OS. Collins equation was used to derive analytical formulas for the electrical field and the effective beam length. These equations could be used to research the propagation of a number of laser beams, such as Gaussian, flat-topped, sine-Gaussian, sinh-Gaussian, cosh-Gaussian, Hermite-sine-Gaussian, Hermite-cosh-Gaussian, Hermite-sinh-Gaussian, Hermite-cos-Gaussian, and higher-order annular Gaussian, using a FrFT operating system. Fourier transformation (FrFT) fractional properties of two mirror resonators have been investigated in 2010 (Moreno et al. 2010). Du et al. (2011) derived approximate theoretical equations for both Lorentz-Gauss and Lorentz beams propagating through an opened FrFT OS. Hashemi et al. (2011) found cosh-square-Gaussian beam properties through an opening and ideal FrFT systems. These studies suggested that the FrFT order of normalized intensity distributions were periodic with a period 2 and independent of the aperture effect. Wang et al. (2011) reported the FrFT coincidence with the stochastic Gaussian Schell (EGSM) model. An analytical expression for Gaussian beams passing through an anamorphic FrFT system with an eccentric circular aperture has been proposed by Zhang et al. (2011). Tang et al. (2012) studied the propagation properties of hypergeometric confluent (HyG) beams propagating through FrFT OSs based on FrFT in the cylindrical coordinate system.

2.4.6 Cryptography

In the literature, few studies reported the cryptography using FrFT and modified FrFT (Ran et al. 2009; Pei and Hsue 2006; Tao et al. 2008b; Youssef 2008). The combined modified FrFT and double random phase encoding technique has been applied for encryption of digital data. The data security has been enhanced with this FrFT-based encryption. Another variant of FrFT called random discrete FrFT is also proposed for cryptography. In random discrete FrFT, both the magnitude and phase transforms are random. Results proved the potentiality of this method for image encryption.

2.4.7 Communications

In telecommunications, orthogonal frequency-division multiplexing (*OFDM*) is a method of encoding digital data on multiple carrier frequencies. The traditional multicarrier (MC) systems are designed based on DFT. However, these methods produce non-stationary results for doubly selective channels (channels with selectivity in both time and frequency). In 2001, Martone et al. proposed discrete FrFT-based multi carrier systems for OFDM (Martone 2001). The proposed scheme outperformed the DFT based method. A minimum mean squared error receiver is proposed for MIMO systems based on the FrFT with space time filtering over fading Rayleigh channels (Khanna and Saxena 2009; Lakshminarayana et al. 2009). The proposed FrFT-based receiver achieved superior performance over the conventional simple mean squared error receiver. The demodulation of multiple access system is developed using the chirp modulation spread spectrum based on the FrFT.

2.5   FrFT for Speech Enhancement Application

The FrFT, which is the generalization form of the classical Fourier transformation, can be used successfully in all cases where the Fourier transformation is actually used. Some of the typical applications of FrFT in the area of signal processing are discussed in the previous section. In this section, some of the FrFT applications in the area of speech enhancement will be presented. For speech enhancement, FrFT has been utilized for optimal filtering, FIR filtering and Spectral subtraction (SS) approaches so far and their results proved the efficacy of the FrFT-based approaches over Fourier transform based methods.

2.5.1   Spectral Subtraction

Wang Zhenli et al. and Ruthwik et al. implemented FrFT-based spectral subtraction (Rutwik and Krishna 2015; Wang and Zhang 2005). The process of fractional spectral subtraction is same processing used in conventional spectral subtraction but substitutes the FFT by FrFT. The schematic diagram for FrFT-based spectral subtraction is shown in Fig. 2.4. In this method, the fractional Fourier transform is applied to the frames of noisy speech samples. The estimated fractional noise spectrum is then subtracted from the derived fractional speech-noise spectrum. Finally, the enhanced speech is obtained from inverse fractional Fourier transform.

$$\text{Let } x(n) = s(n) + d(n) \tag{2.49}$$

where $s(n)$ is the clean speech signal and $d(n)$ is the additive noise. By applying FrFT to the above equation, the following equation can be obtained:

$$X_p(n) = S_p(n) + D_p(n) \tag{2.50}$$

where X_p, S_p, and D_p represent DFrFT of $x(n)$, $s(n)$, and $d(n)$, respectively.

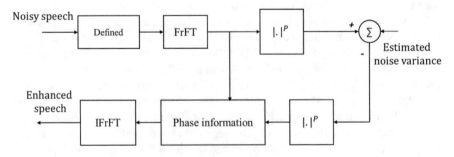

Fig. 2.4   Schematic diagram of signal filtering in the FrFD

When $p = 1$, the above equation reduces to conventional DFT.

Let W_0 and ΔW_0 represent the optimal filter and random perturbation factor, respectively, then,

$$W_p = W_0 + \beta \Delta W_0 \tag{2.51}$$

$$\widehat{S}_p = W_p X_p = (W_0 + \beta \Delta W_0) X_p \tag{2.52}$$

The minimum mean square error is the cost function which is given as

$$J = \sigma^2 = E\left(\left|S_p - \widehat{S}_p\right|^2\right) \tag{2.53}$$

From Eqs. (2.52) and (2.53),

$$\frac{\partial \widehat{S}_p}{\partial \beta} = \frac{\partial}{\partial \beta}(W_p X_p) = \frac{\partial}{\partial \beta}\left[(W_0 + \beta \Delta W_0) X_p\right] = \Delta W_0 X_p \tag{2.54}$$

$$\frac{\partial J}{\partial \beta} = \frac{\partial}{\partial \beta}\left[E\left(\left|S_p - \widehat{S}_p\right|^2\right)\right] = 2\Delta W_0\left(W_p[R_{XX}]_p - [R_{SX}]_p\right) \tag{2.55}$$

$$\left.\frac{\partial J}{\partial \beta}\right|_{\beta=0} = 0 \tag{2.56}$$

W_p can be obtained by equating

$$W_p = \frac{[R_{SX}]_p}{[R_{XX}]_p} \tag{2.57}$$

$$J = \sigma^2 = E\left(\left|S_p - \widehat{S}_p\right|^2\right) \tag{2.58}$$

$$= [R_{SS}]_p - 2W[R_{SX}]_p + \left|W^2\right|[R_{XX}]_p \tag{2.59}$$

From Eq. (2.50)

$$S_p X_p = X_p - D_p \tag{2.60}$$

$$\widehat{S}_p = W_p X_p = (S_p + D_p)W_p = W_p(FrFT(S + N)) \tag{2.61}$$

Time domain value is obtained by taking inverse transform to Eq. (2.61)

$$\widehat{S}(t) = FrFT^{-1}\left(\widehat{S}_p\right) = FrFT^{-1}\left(W_p(FrFT(S + N))\right) \tag{2.62}$$

2.5.2 Optimal Transform for Speech Processing

Optimal transform for speech processing is derived by using the information of pitch, harmonics, and formants of Gammatone filter banks. The authors applied this method for MFCC and speech enhancement based on spectral subtraction (Fig. 2.5).

In this method, the chirp rate of noisy speech is given as

$$\alpha(i,k) = -\frac{1}{2} + a \, \cot\left(2\pi \frac{f_c(i,k)}{P(i)} \delta\right) \tag{2.63}$$

where $P(i)$ is the pitch, $f_c(i,k)$ is the central frequency of kth gammatone filter in ith frame. $\alpha(i,k)$ is the kth candidate transform order in i-8th frame, δ represents pitch rate in frame.

The features based on FrFT are computed with same processing used for MFCC, but substituting the FFT by FrFT. Spectral subtraction in FrFD is also implemented as conventional SS but by replacing the FFT with FrFT.

2.5.3 Filtering

FrFT-based filtering using thresholding for speech enhancement is proposed in (Wang 2014; Ram and Mohanty 2017). The schematic diagram of this method is shown in Fig. 2.6.

In this method, FIR filter is implemented using FrFT.

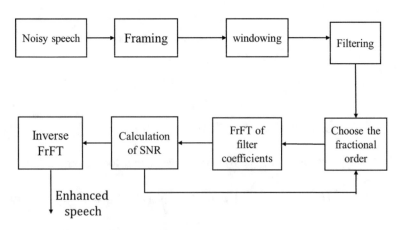

Fig. 2.5 Schematic diagram of signal filtering in the FrFD

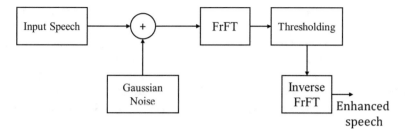

Fig. 2.6   Schematic diagram of FrFT filtering with thresholding

$$h(m) = h_d(m)w(m) \tag{2.64}$$

$$H(\omega) = \frac{1}{2\pi} \int\limits_{-\pi}^{\pi} W(\theta)H_d(\omega - \theta)d\theta \tag{2.65}$$

here, $w(m)$ is a window function, $h_d(m)$ is desired impulse response, and $H_d(\omega - \theta)$ is desired frequency response.

Filter output is defined as

$$y(m) = \int\limits_{-\infty}^{\infty} h(m - \theta)x(\theta)d\theta \tag{2.66}$$

In FFT,

$$y(m) = \frac{1}{\sqrt{2\pi}} \mathrm{IFT}(\mathrm{FT}(x(m)).H(\omega)) \tag{2.67}$$

where $H(\omega) = \mathrm{FT}(h(m))$ for fractional filter,

$$Y(m) = F_{-p}\big\{F_{-p}\{x(m)\}.H_p(\epsilon)\big\} \tag{2.68}$$

$$H_p(\epsilon) = F_p\{h(m)\} \tag{2.69}$$

2.6   Conclusions

FrFT is a generalization of the classical Fourier transform. It is gaining an interesting attention in recent years in the area of signal processing and time-frequency analysis. This chapter attempts to provide an extensive overview of theoretical and computational aspects of continuous and discrete fractional Fourier transforms. There are many ways to implement DFrFT and each method is briefly reviewed along with its advantages and limitations.

Many useful applications of FrFT are mentioned in this chapter. FrFT is a powerful tool for non-stationary signal processing. As the speech signal is non-stationary in nature FrFT has also been adopted for digital speech processing mainly in the areas of speech enhancement and feature extraction. The concepts of spectral subtraction and filtering in FrFD for speech signals is presented in this chapter.

Eigenvector decomposition type-DFrFT has minimum constraints on signals when compared to other types of DFrFT hence it found many applications. It has high computational complexity. Nevertheless, it can be implemented in real time with the parallel processing of high-speed DSP processors. The application of FrFT for two channel and single channel enhancement techniques are described in the following chapters.

References

Almeida, L. B. (1994). The fractional Fourier transform and time-frequency representations. *IEEE Transactions on Signal Processing, 42*(11), 3084–3091. https://doi.org/10.1109/78.330368

Ashok Narayanan, V., & Prabhu, K. M. M. (2003). The fractional Fourier transform: Theory, implementation and error analysis. *Microprocessors and Microsystems, 27*(10), 511–521. https://doi.org/10.1016/S0141-9331(03)00113-3

Barshan, B., & Ayrulu, B. (2002). Fractional Fourier transform pre-processing for neural networks and its application to object recognition. *Neural Networks, 15*(1), 131–140. https://doi.org/10.1016/S0893-6080(01)00120-4

Candan, C., Kutay, M. A., & Ozaktas, H. M. (2000). The discrete fractional Fourier transform. *IEEE Transactions on Signal Processing, 48*(5), 1329–1337. https://doi.org/10.1109/78.839980

Candan, Ç., & Ozaktas, H. M. (2003). Sampling and series expansion theorems for fractional Fourier and other transforms. *Signal Processing, 83*(11), 2455–2457. https://doi.org/10.1016/S0165-1684(03)00196-8

Cariolaro, G., Erseghe, T., Kraniauskas, P., & Laurenti, N. (1998). A unified framework for the fractional Fourier transform. *IEEE Transactions on Signal Processing, 46*(12), 3206–3219. https://doi.org/10.1109/78.735297

Dickinson, B., & Steiglitz, K. (1982). Eigenvectors and functions of the discrete Fourier transform. *IEEE Transactions on Acoustics, Speech, and Signal Processing, 30*(1), 25–31. https://doi.org/10.1109/TASSP.1982.1163843

Du, W., Zhao, C., & Cai, Y. (2011). Propagation of Lorentz and Lorentz–Gauss beams through an apertured fractional Fourier transform optical system. *Optics and Lasers in Engineering, 49*(1), 25–31. https://doi.org/10.1016/j.optlaseng.2010.09.004

Erden, M. F., Kutay, M. A., & Ozaktas, H. M. (1999). Degrees of freedom whereas shift-invariant systems have only. *Electrical Engineering*, 481–485.

Erseghe, T., Kraniauskas, P., & Cariroaro, G. (1999). Unified fractional Fourier transform and sampling theorem. *IEEE Transactions on Signal Processing, 47*(12), 3419–3423. https://doi.org/10.1109/78.806089

Gao, Y.-Q., Zhu, B.-Q., Liu, D.-Z., & Lin, Z.-Q. (2010). Fractional Fourier transform of flat-topped multi-Gaussian beams. *Journal of the Optical Society of America A, 27*(2), 358. https://doi.org/10.1364/JOSAA.27.000358

Guo, Q., Liu, Z., & Liucora, S. (2011). Image watermarking algorithm based on fractional Fourier transform and random phase encoding. *Optics Communications, 284*(16–17), 3918–3923. https://doi.org/10.1016/j.optcom.2011.04.006

Han, D., Liu, C., & Lai, X. (2012). The fractional Fourier transform of Airy beams using Lohmann and quadratic optical systems. *Optics & Laser Technology, 44*(5), 1463–1467. https://doi.org/10.1016/j.optlastec.2011.12.017

Hashemi, S. S., Sabouri, S. G., & Soltanolkotabi, M. (2011). A study of propagation of cosh-squared-Gaussian beam through fractional Fourier transform systems. *Optica Applicata, 41*(4), 897–909.

Jin, X.-W., Lu, M.-F., Xie, Y.-A., & Yu, Z.-X. (2016). Sampled signal analysis in the fractional Fourier transform domain. In *2016 URSI Asia-Pacific Radio Science Conference (URSI AP-RASC)* (pp. 1489–1492). IEEE. https://doi.org/10.1109/URSIAP-RASC.2016.7601170

Khanna, R., & Saxena, R. (2009). Improved fractional Fourier transform based receiver for spatial multiplexed MIMO antenna systems. *Wireless Personal Communications, 50*(4), 563–574. https://doi.org/10.1007/s11277-008-9637-4

Kutay, A., Ozaktas, H. M., Ankan, O., & Onural, L. (1997). Optimal filtering in fractional Fourier domains. *IEEE Transactions on Signal Processing, 45*(5), 1129–1143. https://doi.org/10.1109/78.575688

Lakshminarayana, H. K., Bhat, J. S., Jagadale, B. N., & Mahesh, H. M. (2009). Improved chirp modulation spread spectrum receiver based on fractional Fourier transform for multiple access. In *2009 International Conference on Signal Processing Systems* (pp. 282–286). IEEE. https://doi.org/10.1109/ICSPS.2009.65

Li, Z., Yang, J., Lan, R., & Feng, X. (2012). Multilayer-pseudopolar fractional Fourier transform approach for image registration. In *2012 Eighth International Conference on Computational Intelligence and Security* (pp. 323–327). IEEE. https://doi.org/10.1109/CIS.2012.79

Li, Z., Yang, J., Li, M., & Lan, R. (2013). Estimation of large scalings in images based on multilayer pseudopolar fractional Fourier transform. *Mathematical Problems in Engineering, 2013*, 1–9. https://doi.org/10.1155/2013/179489

Martone, M. (2001). A multicarrier system based on the fractional Fourier transform for time-frequency-selective channels. *IEEE Transactions on Communications, 49*(6), 1011–1020. https://doi.org/10.1109/26.930631

Mendlovic, D., & Ozaktas, H. M. (1993). Fractional Fourier transforms and their optical implementation: I. *Journal of the Optical Society of America A, 10*(9), 1875. https://doi.org/10.1364/JOSAA.10.001875

Moreno, I., Garcia-Martinez, P., & Ferreira, C. (2010). Teaching stable two-mirror resonators through the fractional Fourier transform. *European Journal of Physics, 31*(2), 273–284. https://doi.org/10.1088/0143-0807/31/2/004

O'Neill, J. C., & Williams, W. J. (1996). New properties for discrete, bilinear time-frequency distributions. In *Proceedings of Third International Symposium on Time-Frequency and Time-Scale Analysis (TFTS-96)* (pp. 505–508). IEEE. https://doi.org/10.1109/TFSA.1996.550103

Ozaktas, H. M., Arikan, O., Kutay, M. A., & Bozdagt, G. (1996). Digital computation of the fractional Fourier transform. *IEEE Transactions on Signal Processing, 44*(9), 2141–2150. https://doi.org/10.1109/78.536672

Ozaktas, H. M., Barshan, B., & Mendlovic, D. (1994). Convolution and filtering in fractional Fourier domains. *Optical Review, 1*(1), 15–16. https://doi.org/10.1007/s10043-994-0015-5

Ozaktas, H. M., & Mendlovic, D. (1993). Fractional Fourier transforms and their optical implementation II. *Journal of the Optical Society of America A, 10*(12), 2522. https://doi.org/10.1364/JOSAA.10.002522

Pei, S.-C., & Yeh, M.-H. (1997). Improved discrete fractional Fourier transform. *Optics Letters, 22*(14), 1047. https://doi.org/10.1364/OL.22.001047

Pei, S.-C., & Ding, J.-J. (2000). Closed-form discrete fractional and affine Fourier transforms. *IEEE Transactions on Signal Processing, 48*(5), 1338–1353. https://doi.org/10.1109/78.839981

Pei, S.-C., & Hsue, W.-L. (2006). The multiple-parameter discrete fractional Fourier transform. *IEEE Signal Processing Letters, 13*(6), 329–332. https://doi.org/10.1109/LSP.2006.871721

Pei, S.-C., Tseng, C.-C., Yeh, M.-H., & Shyu, J.-J. (1998). Discrete fractional Hartley and Fourier transforms. *IEEE Transactions on Circuits and Systems II: Analog and Digital Signal Processing, 45*(6), 665–675. https://doi.org/10.1109/82.686685

Pei, S.-C., Yeh, M.-H., & Tseng, C.-C. (1999). Discrete fractional Fourier transform based on orthogonal projections. *IEEE Transactions on Signal Processing, 47*(5), 1335–1348. https://doi.org/10.1109/78.757221

Ram, R., & Mohanty, M. N. (2017). Design of fractional Fourier transform based filter for speech enhancement. *International Journal of Control Theory and Applications, 10*(7), 235–243.

Ran, Q., Zhang, H., Zhang, J., Tan, L., & Ma, J. (2009). Deficiencies of the cryptography based on multiple-parameter fractional Fourier transform. *Optics Letters, 34*(11), 1729. https://doi.org/10.1364/OL.34.001729

Rutwik, A., & Krishna, B. T. (2015). Speech enhancement using fractional order spectral subtraction method. *International Journal of Control Theory and Applications, 8*(3), 1063–1070.

Seok, J., & Bae, K. (2014). Target classification using features based on fractional Fourier transform. *IEICE Transactions on Information and Systems, E97.D*(9), 2518–2521. https://doi.org/10.1587/transinf.2014EDL8003

Sharma, K. K., & Joshi, S. D. (2005). Fractional Fourier transform of bandlimited periodic signals and its sampling theorems. *Optics Communications, 256*(4–6), 272–278. https://doi.org/10.1016/j.optcom.2005.07.003

Singh, K. (2006). *Performance of discrete fractional Fourier transform classes in signal processing applications.* Retrieved from http://tudr.thapar.edu:8080/jspui/bitstream/123456789/94/1/P92233.pdf

Tang, B., Jiang, C., & Zhu, H. (2012). Fractional Fourier transform for confluent hypergeometric beams. *Physics Letters A, 376*(38–39), 2627–2631. https://doi.org/10.1016/j.physleta.2012.07.017

Tant, K. M. M., Mulholland, A. J., Langer, M., & Gachagan, A. (2015). A fractional Fourier transform analysis of the scattering of ultrasonic waves. *Proceedings of the Royal Society A: Mathematical, Physical and Engineering Sciences, 471*(2175), 20140958. https://doi.org/10.1098/rspa.2014.0958

Tao, R., Deng, B., Zhang, W.-Q., & Wang, Y. (2008a). Sampling and sampling rate conversion of band limited signals in the fractional Fourier transform domain. *IEEE Transactions on Signal Processing, 56*(1), 158–171. https://doi.org/10.1109/TSP.2007.901666

Tao, R., Lang, J., & Wang, Y. (2008b). Optical image encryption based on the multiple-parameter fractional Fourier transform. *Optics Letters, 33*(6), 581. https://doi.org/10.1364/OL.33.000581

Wang, F., Zhu, S., Hu, X., & Cai, Y. (2011). Coincidence fractional Fourier transform with a stochastic electromagnetic Gaussian Schell-model beam. *Optics Communications, 284*(22), 5275–5280. https://doi.org/10.1016/j.optcom.2011.08.002

Wang, J. (2014). Speech enhancement research based on fractional Fourier transform. *TELKOMNIKA Indonesian Journal of Electrical Engineering, 12*(12), 576–581. https://doi.org/10.11591/telkomnika.v12i12.6694

Wang, Z., & Zhang, X. (2005). On the application of fractional Fourier transform for enhancing noisy speech. In *MAPE2005: IEEE 2005 International Symposium on Microwave, Antenna, Propagation and EMC Technologies for Wireless Communications, Proceedings.* (Vol. 1(8)) (pp. 289–292).

Xia, X.-G. (1996). On bandlimited signals with fractional Fourier transform. *IEEE Signal Processing Letters, 3*(3), 72–74. https://doi.org/10.1109/97.481159

Youssef, A. M. (2008). On the security of a cryptosystem based on multiple-parameters discrete fractional Fourier transform. *IEEE Signal Processing Letters, 15*, 77–78. https://doi.org/10.1109/LSP.2007.910299

Zayed, A. I. (1996). On the relationship between the Fourier and fractional Fourier transforms. *IEEE Signal Processing Letters, 3*(12), 310–311. https://doi.org/10.1109/97.544785

Zayed, A. I., & García, A. G. (1999). New sampling formulae for the fractional Fourier transform. *Signal Processing, 77*(1), 111–114. https://doi.org/10.1016/S0165-1684(99)00064-X

Zhang, J., Xu, Q., & Lu, X. (2011). Propagation properties of Gaussian beams through the anamorphic fractional Fourier transform system with an eccentric circular aperture. *Optik, 122* (4), 277–280. https://doi.org/10.1016/j.ijleo.2009.11.032

Zhao, C., & Cai, Y. (2010). Propagation of a general-type beam through a truncated fractional Fourier transform optical system. *Journal of the Optical Society of America A, 27*(3), 637. https://doi.org/10.1364/JOSAA.27.000637

Chapter 3
Dual Channel Speech Enhancement Based on Fractional Fourier Transform

3.1 Basics of Adaptive Noise Cancellation

Adaptive noise cancellation (ANC) is a potential noise reduction technique developed by Widrow (Widrow and Stearns 1985), as an alternative to single channel noise reduction algorithms. The block diagram of ANC is shown in Fig. 3.1. ANC uses two inputs, commonly known as the primary input and the reference input. The signal at the primary input $(S + n_0)$ of the system is composed of a desired component S and noise component n_0, whereas the reference input consists of the reference noise signal n which is correlated with noise n_0 and uncorrelated with the signal S. An adaptive filter provides an estimate of the noise in the primary input. The noise estimated is then subtracted from the primary signal. The output of the adaptive noise canceller is used to adjust the coefficients of the adaptive filter. An adaptation algorithm is used in the ANC method to minimize the MSE. Hence, the output of ANC provides the best estimate of the desired signal with minimum MSE. Widrow formulated a simple and excellent algorithm for ANC called least mean square (LMS) algorithm. The two main assumptions of ANC system are as follows:

- The signal and noise components present at the primary input are uncorrelated.
- The reference signal is correlated with the noise in the primary input.

As shown in Fig. 3.1, ANC requires two inputs, primary input is composed of a signal S and noise N and the reference input of noise is N_0. S and N are uncorrelated and N_0 is correlated with N but not with S. The adaptive filter adjusts the coefficients of the impulse response to minimize the MSE and produces an output $\overline{N}$. This noise estimate $(\overline{N})$ is subtracted from the primary signal to get an estimate of the original signal $\hat{S}$ at the output of ANC.

© The Author(s), under exclusive licence to Springer Nature Switzerland AG 2020   51
P. Kunche, N. Manikanthababu, *Fractional Fourier Transform Techniques for Speech Enhancement*, SpringerBriefs in Speech Technology,
https://doi.org/10.1007/978-3-030-42746-7_3

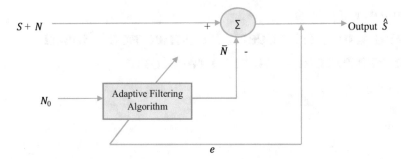

Fig. 3.1 Block diagram of adaptive noise canceller

3.2 Adaptive Filters

Adaptive filter changes the characteristics of the filter in an automated manner to get the best possible output. An adaptive algorithm is used to adjust the parameters of the filter from one iteration to the next. Steepest descent method is a well-known method for adjusting the response of an adaptive system (Hayes 2000; Treichler et al. 1987; Haykin 2001).

3.2.1 LMS

The LMS (Haykin and Widrow 2003) is the prevalent learning algorithm because of its simple realization. The standard LMS algorithm is based on an impulse response filter which minimizes the mean square error (MSE) in a recursive manner, to produce the optimal weights of filter. It is a simplification of steepest descent method, which estimates the gradient vector from the given data. The gradient vector $\nabla J(n)$ is determined by using the following equation:

$$\nabla J(n) = -2p + 2Rw(n) \tag{3.1}$$

where R and p are the estimates of correlation matrix and cross correlation matrix, respectively, and R and p are defined as

$$\widehat{R}(n) = u(n) + u^H(n) \tag{3.2}$$

$$p(n) = u(n) + d^*(n) \tag{3.3}$$

where $u(n)$ is the input vector and $d(n)$ is the vector of the signal to be recovered. Thus we obtain $\nabla J(n)$ as

$$\nabla J(n) = -2u(n) + d^*(n) + 2u(n)u^H(n)w(n) \tag{3.4}$$

The updating vector for steepest decent algorithm is as follows:

$$w(n+1) = w(n) + \mu[p - Rw(n)] \tag{3.5}$$

Substituting Eq. (3.3) for gradient vector in the steepest descent algorithm, we get the following update vector rule for the tap-weight vectors:

$$w(n+1) = w(n) + \mu u(n)\left[d^*(n) - \mu^H(n)\widehat{w}(n)\right] \tag{3.6}$$

where μ is the step size. The error signal $e(n)$ is defined as follows:

$$e(n) = d(n) - y(n) \tag{3.7}$$

The final update equation for the tap weight is given by

$$w(n+1) = w(n) + \mu u(n)e(n) \tag{3.8}$$

3.2.2 NLMS

NLMS (Douglas 1994) is a variant and an improved version of LMS algorithm. NLMS algorithm is developed to overcome the issue of gradient noise amplification in LMS and to obtain higher speed of convergence and more accurate result. The construction of the normalized LMS filter is quite similar to the standard LMS filter, but the manner of updating the weights is different for both. The normalized LMS filter depends on the principle of minimum error. From the one instant to the next, the weight vector of an adaptive filter should be changed in minimizing manner by imposing a minimum constraint on the output of the filter. The rule of adaptation is given by

$$w(n+1) = w(n) + \frac{\widetilde{\mu}}{\delta + \|u(n)\|^2}u(n)e^*(n) \tag{3.9}$$

where $\|u(n)\|^2$ is the total expected energy of the input signal $u(n)$, $\widetilde{\mu}$ is the normalized step size, and δ is a positive scalar that controls the maximum step size μ.

3.3   Application of FrFT Based ANC to Speech Enhancement

3.3.1   Continuous FrFT

The fractional Fourier transform is the generalization of the conventional Fourier transform (FT) and can be interpreted as a counterclockwise rotation of the signal to any angles in the time-frequency plane (Pei and Ding 2003; Ozaktas et al. 2001). The pth order continuous FrFT of a signal $x(t)$ is defined as:

$$F^p[x(t)](u) \equiv X_p(u) = \int_{-\infty}^{\infty} x(t)B_p(t,u)dt \tag{3.10}$$

where $B_p(t, u)$ is the continuous FrFT Kernel given by

$$B_p(t,u) = K_p \exp\left[j\{(t^2 + u^2)/2\} \cot \alpha - jut \operatorname{cosec} \alpha\right] \tag{3.11}$$

and $K_p = \sqrt{(1 - j \cot \alpha)/2\pi}$

The transformation angle $\alpha = p\pi/2$ for $0 < |p| < 2$.

The transformation kernel is also defined as (Pei and Ding 2007a; Pei et al. 2006),

$$B_p(t,u) = \sum_{n=0}^{\infty} e^{-jn\alpha} \varphi_n(t)\varphi_n(u) \tag{3.12}$$

where $\varphi_n(t)$ is the nth order continuous Hermite–Gaussian function. The continuous FrFT of a given signal can be computed as follows:

1. Multiplication by a chirp signal.
2. A Fourier transform with its argument scaled by Cosec.
3. Multiplication by another chirp signal.
4. Multiplication by another constant.

The inverse fractional Fourier transform is defined as (McBride and Kerr 1987; Santhanam and McClellan 1995; Almeida 1994; Pei and Ding 2007b; Pei and Yeh 1997; Namias 1980),

$$x(t) = \int_{-\infty}^{\infty} F^p(u)B_{-p}(t,u)du \tag{3.13}$$

The relationship between fractional domain with the traditional time-frequency plane can be expressed in matrix form as

$$\begin{pmatrix} t \\ w \end{pmatrix} = \begin{pmatrix} \cos \alpha & \sin \alpha \\ -\sin \alpha & \cos \alpha \end{pmatrix} \begin{pmatrix} u \\ v \end{pmatrix} \tag{3.14}$$

Two adaptive algorithms in FRFD are presented. The flowchart for the implementation of FrFT algorithm is shown in Fig. 3.3. Optimal filtering is a very important application of the FrFT. The fractional Fourier domain filtering is a wider class than time invariant Fourier domain filtering. The FrFT can improve the performance of the adaptive filter by reducing the mean square error value (Kutay et al. 1995; Mendlovic et al. 1993; Ozaktas and Mendlovic 1993a, 1995; Mendlovic and Ozaktas 1993; Ozaktas and Mendlovic 1993b; Lohmann 1993). The error coefficients obtained by comparing both the input signals are fed back to the adaptive algorithm to update the coefficients of FrFT-based filter. The updated coefficients of adaptive filter algorithm give the response of desired filter. The analytical framework for LMS and NLMS algorithms in FRFD is presented in following sections.

3.3.2 LMS-FrFT

The block diagram of the adaptive noise reduction algorithm in FRFD is shown in Fig. 3.2, where the inverse fractional Fourier transform is represented as IFrFT. The reference input vector $x(n)$ at time n which contains M adjacent samples and the primary signal $d(n)$ are transformed to fractional Fourier domain expressed as $X_P(n)$ and $D_P(n)$, respectively. The output of the LMS adaptive filter in FRFD is given by

$$Y_P(n) = X(n)W(n) \tag{3.15}$$

where $W(n)$ is the $M \times 1$ vector of the filter coefficients and X $(n) = \text{diag} \{X_{P1}(n), \ldots, X_{Pn}(n)\}$. The difference between desired signal in FRFD $(D_P(n))$ and output of the adaptive filter $(Y_P(n))$ is the error $E_P(n)$ and is given by

$$E_P(n) = D_P(n) - Y_P(n) = D_P(n) - X(n)W(n) \tag{3.16}$$

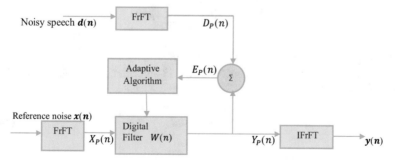

Fig. 3.2 Block diagram of adaptive filtering in FRFD

The new coefficient vector of the filter for the next time interval $W(n + 1)$ can be derived by using the estimate of the ideal cost function and is given by

$$W(n+1) = W(n) + \mu_{LMS}X^*(n)E_P(n) \tag{3.17}$$

where μ_{LMS} is the adaptation step size. The MSE is calculated as

$$\varepsilon(n) = E\left[E_P^H(n)E_P(n)\right]/M \tag{3.18}$$

$$= \left\{E\left[D_P^H(n)D_P(n)\right] + W^H(n)R_{xx}W(n) - 2R_{xd}^H W(n)\right\}/M \tag{3.19}$$

where $R_{xx} = E[X^H(n)X(n)]$ is the autocorrelation matrix estimate from the elements of the input data and $R_{xd} = E[X^H(n)D_P(n)]$ is the cross correlation vector. To obtain the stable convergence of the MSE, the condition for the range of step size of LMS algorithm is as follows:

$$0 < \mu_{LMS} < \frac{2}{Tr[R_{xx}]} \tag{3.20}$$

3.3.3   NLMS-FrFT

The NLMS algorithm in FRFD is realized by normalizing the step size μ as given in the following equation:

$$\mu_{NLMS} = \frac{\mu}{\beta + X_P^H(n)X(n)} \tag{3.21}$$

where $X_P^H(n)X_P(n)$ is the tap input power in FrFT domain and β is a positive constant used to avoid zero values in the denominator.

$$X_P^H(n)X_P(n) = \sum_{i=1}^{N} X_{Pi}^2(n) \tag{3.22}$$

It is supposed here that $X_P^H(n)X_P(n) \approx Tr[R_{XX}]$

To provide the stable convergence for NLMS algorithm, the range of step size is given in Eq. (3.20)

$$0 < \mu < 2 \tag{3.23}$$

The steps involved in the process of NLMS adaptive filter in FRFD are:

(a) Choose the fractional transformation order p in which the FrFT gives the minimum MSE of the output signal.

(b) Calculate the FrFT of input signal $x(n)$ for the fractional order p, i.e., $X_P(n)$, and obtain $D_P(n)$ by using Eq. (3.16). Initialize the weight vector $W(0)$ at time $n = 0$.

(c) Compute the error estimate $E_P(n)$ from Eq. (3.5). Choose the values of step size and the stability parameter. Calculate the weight vector $W(n)$ from Eq. (3.17) using the NLMS algorithm.

(d) Go to the above step and repeat until the minimum MSE is achieved or the maximum number of iterations is attained.

(e) Apply the inverse FrFT (IFrFT) $Y_P(n)$ to get the time domain output response signal $y(n)$.

3.4   Simulation and Analysis

Simulation results of adaptive filtering techniques for speech enhancement are presented and their performance is analyzed. The filtering algorithms used in the simulations include the NLMS with fractional Fourier transform approach (NLMS-FrFT), NLMS with time domain adaptation, LMS with fractional Fourier transform approach (LMS-FrFT), LMS with time domain adaptation. MATLAB programming code has been developed for the computation of adaptive algorithms. In order to evaluate the adaptive filtering algorithms based on FRFD, the clean speech signals recorded with 8 kHz sampling frequency are taken from the data base called NOIZEUS (https://ecs.utdallas.edu/loizou/speech/noizeus/). Noise signal is taken as the white Gaussian noise generated in MATLAB and is added to clean speech to get the noisy speech signal. Then the noisy signals are given as primary input to adaptive filter. The additive white Gaussian noise (AWGN) signal is considered as a reference noise input to the adaptive filter. The FrFT is applied on the noisy speech signal and noise reference signal by assuming the value of fractional coefficient value is 1.014. Then the FrFT transformed signals are given as inputs to the adaptive filter to estimate the noise spectrum. The denoised signal in FrFT domain is again transformed into time domain by using inverse FrFT. Thus the noise reduced signal will be obtained. The flow chart for the computation of FrFT is shown in Fig. 3.3.

In the simulations, the parameters of adaptive filers are chosen as $\mu = 0.5$ for LMS, $\mu = 0.95$ and $\beta = 0.01$ for NLMS algorithm, respectively. In Fig. 3.4, MSE values are shown for different values of transform order p. This figure indicates that the MSE can converge to the minimum value only when the FrFT transform order matches with the signal parameters and the algorithm cannot provide best convergence in the other fractional Fourier domains. From this figure, it can be observed that the algorithm converges to the minimum MSE for the fractional order value $p = 1.014$ ($p = 1.014$th Fractional domain).

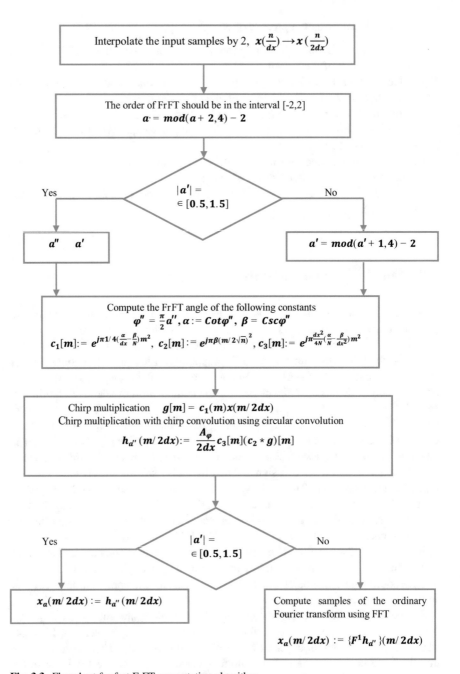

Fig. 3.3 Flowchart for fast FrFT computation algorithm

Fig. 3.4 MSE for different values of transform order p

3.4.1 Parameters of Evaluation

In order to compare the proposed adaptive filtering approach with the other dual channel adaptive filtering algorithms "LMS" and "NLMS" method, four different performance metrics are used. To evaluate the noise reduction, three measures the signal to noise ratio, the peak signal to noise ratio, and the mean square error are used. To evaluate the intelligibility of the enhanced speech, signal to noise ratio loss (SNR$_{\text{LOSS}}$) is used.

SNR
The SNR is expressed as the ratio of the power of filtered signal and the power of noise signal, given by Eq. (3.21). The difference between the SNR of enhanced signal and that of noisy signal is the measure of the improved SNR (SNRI).

$$\text{SNR} = 10\log\left(\frac{P_s}{P_w}\right) \tag{3.21}$$

where P_s is the mean power of the input signal and P_w is the mean power of the noise.

PSNR
Peak signal to noise ratio is defined as the power ratio between a signal and the background noise. PSNR is most important performance parameter in speech enhancement Field. It is measured in dB and it is represented as PSNR. As noise level increases, PSNR value decreases and vice versa. Noise level and PSNR ratio is reciprocal to each other. PSNR is defined as

$$\text{PSNR} = 10 \log 10 \left\{ \frac{\max (A)}{\text{MSE}} \right\} \tag{3.22}$$

where A is the original/clean speech signal.

MSE

Another metric is the Mean Square Error (MSE), which is defined as

$$\text{MSE} = \frac{1}{L} \sum_{k=0}^{N} (D_P(k) - Y_P(k))^2 \tag{3.23}$$

where L is the length of the signal, N is the total number of runs, and $Y_P(k)$ is the output of the filter.

SNR$_{\text{LOSS}}$

The SNR loss (Ma et al. 2011) in band j and frame m is defined as follows:

$$L(j.m) = \text{SNR}_X(j,m) - \text{SNR}_{\widehat{X}}(j,m) \tag{3.24}$$

$$= 10 \log_{10} \frac{X(j,m)^2}{D(j,m)^2} - 10 \log_{10} \frac{\widehat{X}(j,m)^2}{D(j,m)^2} \tag{3.25}$$

$$= 10 \log_{10} \frac{X(j,m)^2}{\widehat{X}(j,m)^2} \tag{3.26}$$

where $\text{SNR}_X(j,m)$ is the input SNR in band j, $SNR_{\widehat{X}}(j,m)$ is the effective SNR of the enhanced signal in the jth frequency band, and $\widehat{X}(j,m)$ is the excitation spectrum of the processed (enhanced) signal in the jth frequency band at the jth frame. The first SNR term in Eq. (3.23) provides the original SNR in frequency band j before processing the input signal $x(n)$, while the second SNR term provides the SNR of the processed (enhanced) signal. The term $L(j, m)$ in Eq. (3.23) thus defines the loss in SNR, termed SNR$_{\text{LOSS}}$, incurred when the corrupted signal goes through a noise-suppression system. Clearly, when $\widehat{X}(j,m) = X(j,m)$, the SNR$_{\text{LOSS}}$ is zero. It is reasonable to expect with most noise-suppression algorithms that as the SNR level increases, i.e., SNR $\to \infty$, the estimated spectrum $\widehat{X}(j,m)$ approaches the clean spectrum $X(j,m)$.

Following the computation of the SNR loss in Eq. (3.23), the $L(j, m)$ term is limited to a range of SNR levels. In the SII index (ANSI 1997), for instance, the SNR calculation is limited to the range of $[-15, 15]$ dB, prior to the mapping of the computed SNR to the range of $[0, 1]$. Assuming in general the restricted SNR range of $[-\text{SNR}_{\text{Lim}}, \text{SNR}_{\text{Lim}}]$ dB, the $L(j, m)$ term is limited as follows:

$$\widehat{L}(j,m) = \min\left(\max\left(L(j,m), -\text{SNR}_{\text{Lim}}\right), \text{SNR}_{\text{Lim}}\right) \tag{3.27}$$

and subsequently mapped to the range of [0, 1] using the following equation:

$$\text{SNR}_{\text{Loss}}(j,m) = \begin{cases} \dfrac{-C_-}{\text{SNR}_{\text{Lim}}}\widehat{L}(j,m) & \text{if } \widehat{L}(j,m) < 0 \\[2mm] \dfrac{C_+}{\text{SNR}_{\text{Lim}}}\widehat{L}(j,m) & \text{if } \widehat{L}(j,m) \geq 0 \end{cases} \tag{3.28}$$

where C_+ and C_- are parameters (defined in the range of [0, 1]) controlling the slopes of the mapping function. The average SNR_{LOSS} is finally computed by averaging $\text{SNR}_{\text{Loss}}(j,m)$ over all frames in the signal as follows:

$$\overline{\text{SNR}}_{\text{Loss}} = \frac{1}{M}\sum_{m=0}^{M-1} f\text{SNR}_{\text{Loss}}(m) \tag{3.29}$$

where M is the total number of data segments in the signal and $f\text{SNR}_{\text{Loss}}(m)$ is the average (across bands) SNR loss computed as follows:

$$f\text{SNR}_{\text{Loss}}(m) = \frac{\sum\limits_{j=1}^{K} W(j)\, SNR_{Loss}(j,m)}{\sum\limits_{j=1}^{K} W(j)} \tag{3.30}$$

where $W(j)$ is the weight (i.e., band importance function (ANSI 1997)) placed on the jth frequency band. Based on Eq. (3.28), it is easy to show that Eq. (3.30) can be decomposed into two terms as follows (assuming for convenience that $W(j) = 1$ for all j):

$$f\text{SNR}_{\text{Loss}}(m) = \frac{1}{K}\left[\sum_{j:L(j,m)\geq 0} \text{SNR}_{\text{Loss}}(j,m) + \sum_{j:L(j,m)<0} \text{SNR}_{\text{Loss}}(j,m)\right] \tag{3.31}$$

3.4.2   Performance Analysis

In Fig. 3.5, speech spectrograms are presented for comparison between the clean signal, the noisy speech input and the output enhanced signals of the studied FrFT-based filtering and time domain adaptive filtering methods. From these

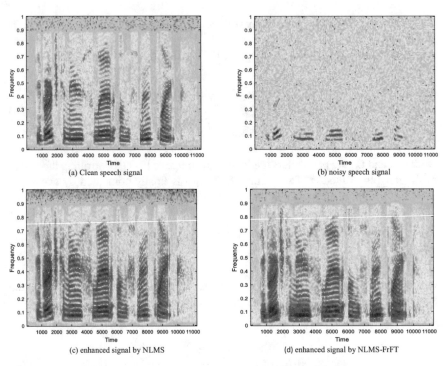

Fig. 3.5 Spectrograms of the (**a**) clean speech signal, (**b**) noisy speech (noise added at 0 dB), (**c**) signal enhanced by NLMS, and (**d**) signal enhanced by NLMS-FrFT filter

Table 3.1 Objective measures SNR, PSNR, and MSE of NLMS and NLMS-FrFT for different input SNR levels

	SNR		PSNR		MSE	
Input SNR (dB)	NLMS	NLMS-FrFT	NLMS	NLMS-FrFT	NLMS	NLMS-FrFT
−5	6.8898	7.7931	7.7688	8.8115	0.0672	0.0443
0	8.0888	11.3081	7.9138	9.0863	0.0452	0.0360
5	9.2150	11.9003	8.8507	9.4211	0.0371	0.0249
10	11.4065	13.3651	8.1694	8.3983	0.0357	0.0229

spectrograms it can be clearly seen that FrFT-based NLMS filter provided better noise reduction when compared to conventional NLMS adaptive filter.

The SNR, PSNR, MSE results, averaged over 10 trail runs, are shown in Table 3.1, for the time domain NLMS algorithm and FrFT domain NLMS algorithm. High values of the SNR and PSNR denote better performance, whereas low values of MSE indicate better performance of adaptive filter. The graphical representation for the improvement in SNR, PSNR, and MSE are shown in Figs. 3.6, 3.7, and 3.8, respectively. From these figures and the results in Table 3.1, it is noted that NLMS_FrFT performed better than NLMS filter for all the input SNR conditions (−5, 0, 5 and 10 dB). Table 3.2 presents the comparison of SNR, PSNR,

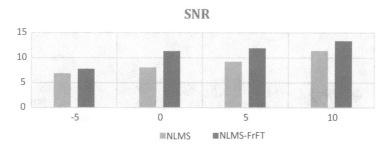

Fig. 3.6 Graphical representation of performance of SNR for NLMS and NLMS-FrFT

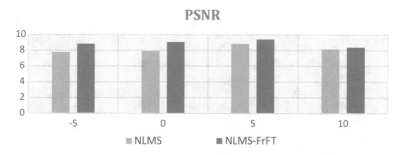

Fig. 3.7 Graphical representation of performance of PSNR for NLMS and NLMS-FrFT

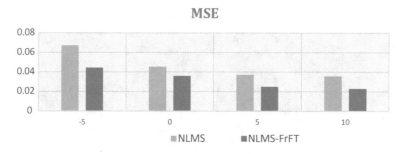

Fig. 3.8 Graphical representation of performance of MSE for NLMS and NLMS-FrFT

Table 3.2 Objective measures SNR, PSNR, and MSE of LMS and LMS-FrFT for different input SNR

Input SNR (dB)	SNR		PSNR		MSE	
	LMS	LMS-FrFT	LMS	LMS-FrFT	LMS	LMS-FrFT
−5	7.0475	7.4256	6.5082	8.4087	0.0687	0.0447
0	10.4013	10.9083	7.5994	11.3249	0.0461	0.0311
5	10.6266	13.6532	8.0558	8.6931	0.0385	0.0251
10	14.3801	15.3752	7.4889	7.9644	0.0339	0.0231

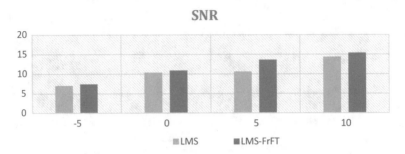

Fig. 3.9 Graphical representation of performance of SNR for LMS and LMS-FrFT

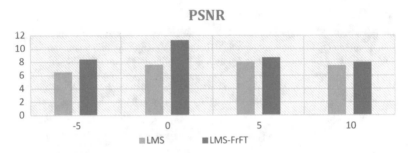

Fig. 3.10 Graphical representation of performance of PSNR for LMS and LMS-FrFT

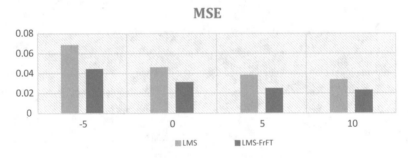

Fig. 3.11 Graphical representation of performance of MSE for LMs and LMS-FrFT

and MSE for the LMS adaptive algorithm and FrFT-based LMS adaptive algorithm and the corresponding graphical representations are shown in Figs. 3.9, 3.10, and 3.11, respectively. From these results it can be clearly observed that LMS algorithm in FrFT domain outperformed the LMS in time domain throughout the input SNR levels.

Figure 3.12 shows the comparison for the time domain waveforms of the speech signals. Figure 3.12a–d shows the original signal, corrupted signal at 0 dB input SNR, signal processed by proposed NLMS-FrFT and signal processed by the conventional NLMS. Comparing Fig. 3.12c and d, it can be noticed that

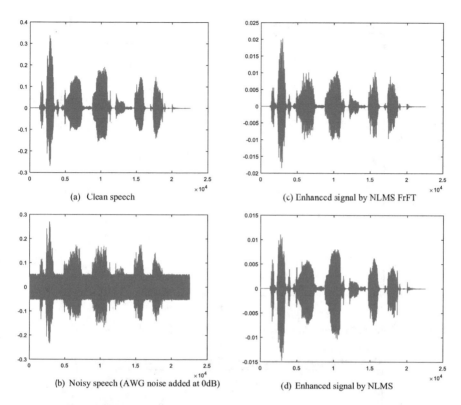

Fig. 3.12 Time domain waveforms of the signals (**a**) clean speech, (**b**) noisy speech, (**c**) signal enhanced by NLMs-FrFT, and (**d**) signal enhanced by NLMS

the proposed method gives a cleaner result than NLMS. The comparison of the estimated and actual weights of the LMS filter under 5 dB input noise level condition is presented in Fig. 3.13. Figure 3.14 provides the comparison of the estimated and actual weights of the LMS-FrFT filter under 5 dB input noise level condition. From these figures it can be inferred that FrFT-based LMS adaptive filter estimates closely to the actual weights. In Figs. 3.15 and 3.16, the comparison of actual weights and estimated weights for NLMS and NLMS_FrFT under 0 dB input noise level are presented, respectively. From the two figures it can be clearly observed that NLMS-FrFT gives performance in terms of noise estimation. Figure 3.17 illustrates the spectrograms of the enhanced speech signal for different values of transform order p. From this figure it is noted that the spectrogram of speech enhanced by NLMS_FrFT with transform order $p = 1.014$ closely resembles the spectrogram of clean speech. To analyze the performance of the algorithms in terms of intelligibility, SNR_{LOSS} of algorithms is compared in Fig. 3.18. It clearly shows the significantly improved performance of SNR_{LOSS} measure under all input noise levels.

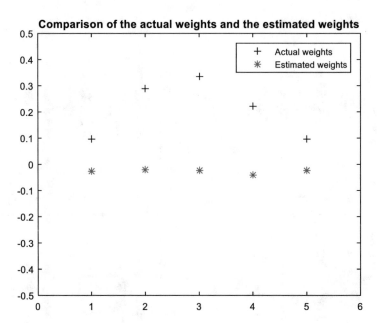

Fig. 3.13 Comparison of the actual and estimated weights of LMS algorithm

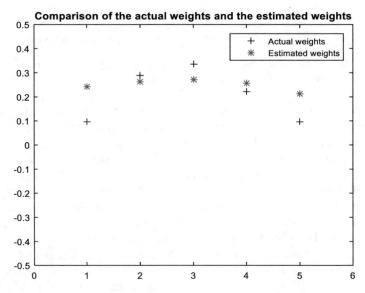

Fig. 3.14 Comparison of the actual and estimated weights of LMS-FrFT algorithm

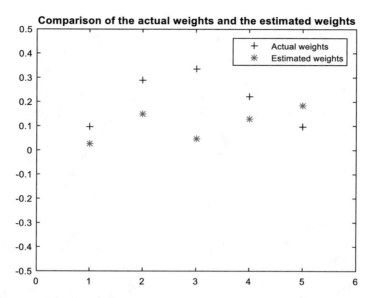

Fig. 3.15 Comparison of the actual and estimated weights of NLMS algorithm

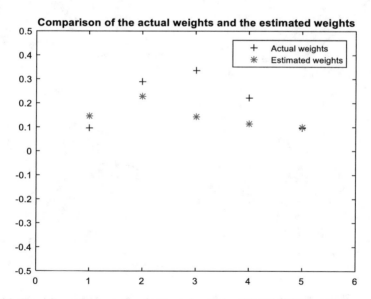

Fig. 3.16 Comparison of the actual and estimated weights of NLMS-FrFT algorithm

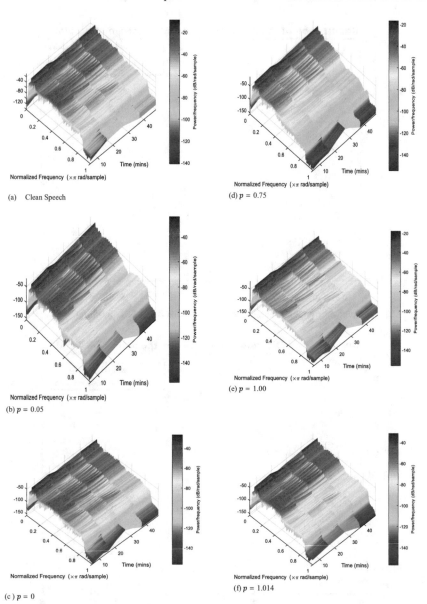

Fig. 3.17 Spectrograms of the signals (**a**) clean speech, (**b**) enhanced speech for $p = 0.05$, (**c**) enhanced signal for $p=0$, (**d**) enhanced signal for $p=0.75$, (**e**) enhanced signal for $p= 1.00$, and (**f**) enhanced signal for $p= 1.014$

To summarize, the simulation results have clearly shown that the performance of fractional Fourier transform based adaptive noise cancellation is much better for reducing speech signal distortions in comparison with adaptive filters in time domain. Moreover, NLMS_ FrFT algorithm based adaptive filtering outperformed LMS-FrFT adaptive algorithms in terms of SNR and MSE convergence.

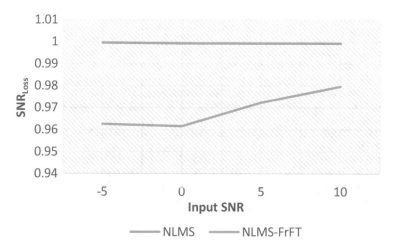

Fig. 3.18 SNR_{LOSS} for different input SNR

3.5 Conclusions

This chapter apprises the fractional Fourier transform based adaptive filtering scheme for speech signal enhancement application to reduce the noise contaminated in speech signals.

- The noisy signal is rotated on time-frequency plane to extract the signal in FRFD. Two adaptive algorithms called LMS and NLMS algorithms are implemented in fractional Fourier domain.
- Evaluation of the proposed algorithm is carried out by computing four objective measures. Fractional Fourier domain adaptive techniques reduced the mean square error when compared to classical time domain algorithms.
- In comparison to the reference adaptive filters LMS and NLMS, the FrFT-based filters LMS-FrFT and NLMS-FrFT yielded a consistently higher noise attenuation together with less speech distortion by obtaining high SNR, PSNR, and low SNR_{LOSS}.
- rt from the quantitative evaluation, both the visual inspection of the speech waveforms and spectrograms, verified the potential of the fractional Fourier transform based adaptive filtering techniques as a robust dual channel enhancement approach.

References

Almeida, L. B. (1994). The fractional Fourier transform and time-frequency representations. *IEEE Transactions on Signal Processing, 42*(11), 3084–3091.

ANSI. (1997). *Methods for calculation of the speech intelligibility index.* Technical report S3.5-1997. New York: American National Standards Institute.

Douglas, S. C. (1994). A family of normalized LMS algorithms. *IEEE Signal Processing Letters, 1*(3), 49.

Hayes, M. H. (2000). *Statistical digital signal processing and modeling.* New York: Wiley. ISBN: 0-471-59431-8.

Haykin, S. (2001). *Adaptive filter theory* (4th ed.). Upper Saddle River: Prentice Hall.

Haykin, S. S., & Widrow, B. (Eds.). (2003). *Least-mean-square adaptive filters.* Hoboken: Wiley.

Kutay, M. A., Ozaktas, H. M., Onural, L., & Arıkan, O. (1995). Optimal filtering in fractional Fourier domains. In *Proceedings of the IEEE International Conference on Acoustics Speech and Signal Processing* (pp. 937–940).

Lohmann, A. W. (1993). Image rotation, Wigner rotation and the fractional Fourier transform. *Journal of the Optical Society of America A, 10*, 2181–2186.

Ma, J., Loizou, P. C., & Loss, S. N. R. (2011). A new objective measure for predicting speech intelligibility of noise-suppressed speech. *Speech Communication, 53*(3), 340–354.

McBride, A. C., & Kerr, F. H. (1987). On namias' fractional Fourier transforms. *IMA Journal of Applied Mathematics, 39*, 159–175.

Mendlovic, D., & Ozaktas, H. M. (1993). Fractional Fourier transformations and their optical implementation: Part I. *Journal of the Optical Society of America A, 10*, 1875–1881.

Mendlovic, D., Ozaktas, H. M., & Lohmann, A. W. (1993). Fourier transforms of fractional order and their optical interpretation. In *Proceedings of the Topical Meeting on Optical Computing, OSA Technical Digest Series*, Washington, DC (pp. 127–130).

Namias, V. (1980). The fractional order Fourier transform and its application to quantum mechanics. *IMA Journal of Applied Mathematics, 25*, 241–265.

Ozaktas, H. M., & Mendlovic, D. (1993a). Fourier transforms of fractional order and their optical interpretation. *Optics Communication, 101*, 163–169.

Ozaktas, H. M., & Mendlovic, D. (1993b). Fractional Fourier transformations and their optical implementation: Part II. *Journal of the Optical Society of America A, 10*, 2522–2531.

Ozaktas, H. M., & Mendlovic, D. (1995). Fractional Fourier optics. *Journal of the Optical Society of America A, 12*, 743–751.

Ozaktas, H. M., Zalevsky, Z., & Kutay, M. A. (2001). *The fractional Fourier transform.* Chichester: Wiley.

Pei, S. C., & Ding, J. J. (2003). Eigenfunctions of the offset Fourier, fractional Fourier, and linear canonical transforms. *Journal of the Optical Society of America A, 20*(3), 522–532.

Pei, S. C., & Ding, J. J. (2007a). Relations between Gabor transforms and fractional Fourier transforms and their applications for signal processing. *IEEE Transactions on Signal Processing, 55*(10), 4839–4850.

Pei, S. C., & Ding, J. J. (2007b). Eigen functions of Fourier and fractional Fourier transforms with complex offsets and parameters. *IEEE Transactions on Signal Processing, 54*(7), 1599–1611.

Pei, S. C., Hsue, W. L., & Ding, J. J. (2006). Discrete fractional Fourier transform based on new nearly tridiagonal commuting matrices. *IEEE Transactions on Signal Processing, 54*(10), 3815–3828.

Pei, S. C., & Yeh, M. H. (1997). Improved discrete fractional Fourier transform. *Optics Letters, 22*, 1047–1049.

Santhanam, B., & McClellan, J. H. (1995). The DRFT—A rotation in time frequency space. In *Proceedings of the ICASSP* (pp. 921–924).

Treichler, J. R., Johnson, C. R., & Larimore, M. G. (1987). *Theory and design of adaptive filters.* New York: Wiley.

Widrow, B., & Stearns, S. (1985). *Adaptive signal processing.* Englewood Cliffs, NJ: Prentice Hall.

Chapter 4
Fractional Cosine Transform Based Single Channel Speech Enhancement Techniques

4.1 Fundamentals of Discrete Fractional Cosine Transform

S. C. Pei et al. proposed discrete fractional cosine transform DFrCT in 2001 (Pei and Yeh 2001). DFrCT is a generalization of discrete cosine transform (DCT). The definition of DFrCT is derived based on the Eigen decomposition of DCT, similar to DFrFT. The computation of DFrFT for even signals can be planted into half size DFrFT.

The eigenvectors and eigenvalues of DFT kernel matrix are discussed in Chap. 2. DFrCT is implemented by using the DFT Hermite eigenvectors and DCT kernel.

The DCT has four types of kernel matrices as follows:

DCT-I

$$C_{N+1}^{I} = \sqrt{\frac{2}{N}} \left[k_m k_n \cos \left(\frac{mn\pi}{N} \right) \right]$$

(4.1)

for $m, n = 0, 1, \ldots N$

DCT-II

$$C_{N+1}^{II} = \sqrt{\frac{2}{N}} \left[k_m \cos \left(m \frac{\left(n + \frac{1}{2} \right)\pi}{N} \right) \right]$$

(4.2)

for $m, n = 0, 1, \ldots N$

DCT-III

$$C_{N+1}^{III} = \sqrt{\frac{2}{N}} \left[k_n \cos \left(\frac{\left(m + \frac{1}{2} \right)n\pi}{N} \right) \right]$$

(4.3)

for $m, n = 0, 1, \ldots N$

DCT-IV

$$C_{N+1}^{IV} = \sqrt{\frac{2}{N}} \left[\cos\left(\left(m + \frac{1}{2} \right) \left(\frac{\left(n + \frac{1}{2} \right)\pi}{N} \right) \right) \right]$$

(4.4)

for $m, n = 0, 1, \ldots N$

where $k_m = \begin{cases} 1/\sqrt{2}, & m = 0 \text{ and } m = N \\ 1, & \text{others} \end{cases}$

DCT-I kernel has the properties of symmetry and periodicity. Hence DCT-I is used for derivation of DFrCT kernel matrices. The DCT-I kernel chosen for DFrCT is given as

$$C_N = \sqrt{\frac{2}{N-1}}$$

$$\times \begin{bmatrix} \frac{1}{2} & \frac{1}{\sqrt{2}} & \cdots\cdots & \frac{1}{\sqrt{2}} & \frac{1}{2} \\ \frac{1}{\sqrt{2}} & \cos\frac{\pi}{N-1} & \cdots\cdots & \cos\frac{\pi(N-2)}{N-1} & \frac{1}{\sqrt{2}}\cos\frac{\pi(N-1)}{N-1} \\ \vdots & \vdots & \ddots\cdots & \cdots & \vdots \\ \vdots & \vdots & \vdots\ddots & \cdots & \vdots \\ \frac{1}{\sqrt{2}} & \cos\frac{\pi(N-2)}{N-1} & \vdots\cdots & \cos\frac{\pi(N-2)^2}{N-1} & \frac{1}{\sqrt{2}}\cos\frac{\pi(N-1)(N-2)}{N-1} \\ \frac{1}{2} & \frac{1}{\sqrt{2}}\cos\frac{\pi(N-1)}{N-1} & \vdots\cdots & \frac{1}{\sqrt{2}}\cos\frac{\pi(N-1)(N-2)}{N-1} & \cos\frac{\pi(N-1)^2}{N-1} \end{bmatrix}$$

(4.5)

The eigenvalue multiplicities of DCT-I kernel matrices are given in Table 4.1. The discretization of DFrCT is described as follows:

Let $X_\alpha(u)$ is the FrFT then FrCT is defined as

$$C_\alpha(u) = X_\alpha(u) + X_\alpha(-u)$$

(4.6)

$$= 2\sqrt{\frac{1 - j\cot\alpha}{2\pi}} \times \int_{-\infty}^{\infty} e^{\frac{j\cot\alpha(u^2+t^2)}{2}} \cos\left(utcsc\alpha\right)x(t)dt$$

(4.7)

The samples of $C_\alpha(u)$ can be given as

Table 4.1 Eigenvalue multiplicities of the DCT-I kernel matrices

N	Multiplicity of 1	Multiplicity of -1
Odd	$\frac{N+1}{2}$	$\frac{N-1}{2}$
Even	$\frac{N}{2}$	$\frac{N}{2}$

$$Y_\alpha^c(m) = \sum_{n=0}^{N-1} F_\alpha^c(m,n)y(n), \quad \begin{aligned} m &= 0,1,2\ldots M-1 \\ n &= 0,1,2\ldots N-1 \end{aligned} \tag{4.8}$$

where $u = m\Delta u$ and $t = n\Delta t$, $y(n)$ represents the samples of input with $n = (0, N-1)$.

The kernel $F_\alpha^c(m,n)$ is defined as

$$F_\alpha^c(m,n) = 2\sqrt{\frac{1-j\cot\alpha}{2\pi}}\Delta t \times e^{j\cot\alpha\left(m^2\Delta u^2 + n^2\Delta t^2\right)/2} \times \cos(mn)\Delta t\Delta u csc\alpha \tag{4.9}$$

Inverse DFrCT is defined as

$$y(n) = \sum_{m=0}^{M} F_\alpha^{c*}(m,n)Y_\alpha^c(m) \tag{4.10}$$

where $F_\alpha^{c*}(m,n)$ is a complex conjugate of $F_\alpha^c(m,n)$. From Eqs. (4.8) and (4.9),

$$y(n) = \frac{(2\Delta t)^2}{2\pi|\sin\alpha|}\sum_{k=0}^{N-1} y(k)e^{\frac{j\cot\alpha}{2}(k^2-n^2)\Delta t^2} \times \sum_{m=0}^{M}$$
$$\times (\cos(mn\Delta t\Delta u csc\alpha)\cos(mk\Delta t\Delta u csc\alpha)) \tag{4.11}$$

The kernel of conventional DCT-I is given by

$$C_{N+1} = \sqrt{2/N} \times k_m k_n \cos\left(\frac{mn\pi}{N}\right), m,n = (0,N) \tag{4.12}$$

Hence to get the same kernel, when $\alpha = \frac{\pi}{2}$, sampling should be such that

$$\Delta t\Delta u = \frac{S\pi\sin\alpha}{N} \tag{4.13}$$

where

$$S = \text{sgn}(\sin(\alpha)) \tag{4.14}$$

From Eqs. (4.14) and (4.11),

$$y(n) = \frac{(2\Delta t)^2}{2\pi|\sin\alpha|}\sum_{k=0}^{N-1} y(k)e^{\frac{j\cot\alpha(k^2-n^2)\Delta t^2}{2}} \times \sum_{m=0}^{M} \left(\cos\left(\frac{Smn\pi}{N}\right) \times \left(\frac{Smk\pi}{N}\right)\right) \tag{4.15}$$

To reduce RHS of above equation, to $y(n)$, $F_\alpha^c(m,n)$ should be normalized and then kernel is given as

$$F_\alpha^c(m,n) = k_m k_n \sqrt{\frac{2(1 - j\cot\alpha)|\sin\alpha|}{N}} \times e^{j\cot\alpha(m^2\Delta u^2 + n^2\Delta t^2)/2} \cos\left(\frac{mn\pi}{N}\right) \quad (4.16)$$

where $k_{m,n} = \sqrt{\frac{1}{2}}$ for $m, n = 0,$

$\qquad\qquad = 1$ otherwise

The kernel in equation reduces to that of DCT-I when $\alpha = \frac{\pi}{2}$. To compute IDFrCT, DFrCt is computed with order $-\alpha$, sampling interval Δu at input and Δt at output.

4.1.1 Properties of DFrCT

1. Unitary: The DFrCT is unitary.

$$C_{N,\alpha}^* = C_{N,\alpha}^{-1} = C_{N,-\alpha}$$

2. Angle additive: The DFrCT satisfies the property of angle additivity similar to DFrFT.

$$C_{N,\alpha}\, C_{N,\beta} = C_{N,\alpha+\beta}$$

3. Periodic: The DFrFT is periodic with period 2π whereas DFrCT is periodic with period π.

$$C_{N,\alpha+\pi} = C_{N,\alpha}.$$

4. Symmetric: The DFrCT kernel holds the symmetry property.

$$C_{N,\alpha}(a,b) = C_{N,\alpha}(b,a)$$

The relation between DFrFT and DFrCT is, DFrFT of an even or odd signal can be computed by a smaller-size transform kernel of the DFRCT and DFrST (discrete fractional sine transform). The DFRCT and the DFRST are from the truncated and scaled DFT eigenvectors by (4.15) and (4.17). The definition and derivation for DFrST are given in Chap. 5. The relation between DFrFT and DFrCT is illustrated in Fig. 4.1. Figure 4.1a is the original impulse signal considered, Fig. 4.1b is the DFrFT of original signal, Fig. 4.1c is the half truncation of the impulse signal, and Fig. 4.1d is the DFrCT results of the truncated signal. From this figure, it is observed that the

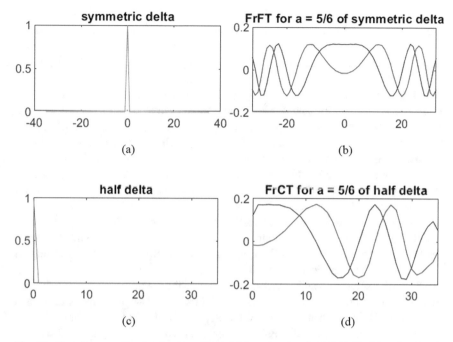

Fig. 4.1 Transform results of an impulse signal. (**a**) Impulse signal. (**b**) DFrFT of impulse signal. (**c**) Half truncation of impulse signal. (**d**) DFrCT of half truncated signal

DFrCT of a truncated signal is equal to the real part of the DFrFT result of the original impulse signal.

4.1.2 Advantages of DFrCT

1. Time varying signal processing can be done effectively with DFrCT.
2. The parameter of angle of rotation or order of the DFrCT transformation can provide an additional degree of freedom which can result in an improved performance over conventional DCT.
3. The DFrFT can be computed using DFrCT and discrete fractional sine transform (DFrST) and hence the computation load of DFrFT can be reduced.

4.1.3 Applications of DFrCT

The fingerprint templates are generated based on the ID discrete fractional Fourier, cosine, and sine transforms (DFrFT, DFrCT, and DFrST) (Yoshimura 2014). These new fingerprint templates realized the fingerprint recognition system with high recognition accuracy and high robustness against attacks. A hybrid discrete fractional cosine transform (DFrCT) with Tikhonov regularization based Turbo minimum mean square error (MMSE) equalization (DFrCT-Turbo) is proposed to suppress inter-carrier interference (ICI) for underwater acoustic channels (UWA) (Chen et al. 2015). The scheme is based on orthogonal frequency division multiplex (OFDM) technique. Moreover, an optimal order selecting method for DFrCT is implemented by maximizing carrier to interference ratio (CIR) to UWA channel character. Results show that proposed method got improvement in BER over traditional orthogonal based methods. The DFrCT is also proposed for watermarking to exactly reconstruct the host image. Results proved that the DFrCT based watermarking is superior to the QIM based watermarking for the biomedical image applications (Ko et al. 2012). The two variants of DFrCT called "three cycles discrete fractional cosine transform" (FDCT3), and modified transform (MFDCT3) are proposed for acoustic emission (AE) signal enhancement (Bultheel and Martínez-Sulbaran 2006). Another variant of DFrCT called reality-preserving fractional discrete cosine transform (RPFrDCT) and spatiotemporal chaotic mapping is presented for the application of color image encryption scheme (Ya-ru et al. 2015). A robust face recognition model which uses the feature extraction capabilities of fractional discrete cosine transform (FDCT) is proposed in Thippeswamy and Shekar (2011). In this work, the DFrCT is applied for the face image database, and the transform coefficients of DFrCT are used as features. The fractional discrete cosine and sine transforms with four random parameters are defined, to which we refer as random fractional discrete cosine and sine transforms. An encryption of color snapshot making use of reality-preserving fractional discrete cosine transform (RPFrDCT) is also reported (Bhagyashri Pandurangi et al. 2017).

4.2   Wiener Filter with Harmonic Regeneration Noise Reduction (W-HRNR)

General additive noise model is represented by

$$p(t) = q(t) + r(t) \tag{4.17}$$

where $q(t)$ is the original/clean signal, $r(t)$ is the noise signal, and $p(t)$ is the noisy signal. Let $Q(f,k)$, $R(f,k)$, and $P(f,k)$ represent the kth spectral component of the short time frame f of the original signal, noise signal, and noisy signal, respectively. To find the spectral estimation of original signal $\widehat{P}(f,k)$, the SNR estimate from noisy features is used. An estimate of $P(f,k)$ is obtained by applying a spectral gain $G(f,k)$ to each short time spectral component $P(f,k)$.

Two important parameters of estimation algorithms are: a posteriori SNR and a priori SNR, respectively, defined by

$$\text{SNR}_{\text{Post}}(f,k) = \frac{|P(f,k)|^2}{E\left[|R(f,k)|^2\right]} \tag{4.18}$$

and

$$\text{SNR}_{\text{Prio}}(f,k) = \frac{E[|Q(f,k)|]^2}{E\left[|R(f,k)|^2\right]} \tag{4.19}$$

where $E[\cdot]$ is the expectation operator. Another parameter called instantaneous SNR is defined by

$$\text{SNR}_{\text{Inst}}(f,k) = \frac{[P(f,k)]^2 E\left[|R(f,k)|^2\right]}{E\left[|R(f,k)|^2\right]} \tag{4.20}$$

$$= \text{SNR}_{\text{Post}}(f,k) - 1 \tag{4.21}$$

Spectral gain $G(f,k)$ is obtained by the function

$$G(f,k) = g\left(\widehat{\text{SNR}}_{\text{Prio}}(f,k), \widehat{\text{SNR}}_{\text{Post}}(f,k)\right) \tag{4.22}$$

depending on chosen distortion measure.

The function g can be chosen among the different gain functions proposed in the literature. Decision direct method is used to estimate the noise based on voice activity detection. Using the obtained noise PSD, the "a posteriori SNR" and the "a priori SNR" are computed as follows:

$$\widehat{\text{SNR}}_{\text{Post}}(f,k) = \frac{|P(f,k)|^2}{\widehat{\gamma}_r(f,k)} \tag{4.23}$$

and

$$\widehat{\text{SNR}}_{\text{Prio}}^{\text{DD}}(f,k) = \beta \frac{\left|\widehat{Q}(f-1.k)\right|^2}{\widehat{\gamma}_r(f,k)} + (1-\beta)D\left[\widehat{\text{SNR}}_{\text{Post}}(f,k) - 1\right] \tag{4.24}$$

where $D[\cdot]$ denotes the half wave rectification and $\widehat{Q}(f-1.k)$ is the previous frame estimated speech spectrum. The $\widehat{\text{SNR}}_{\text{Prio}}^{\text{DD}}(f,k)$ corresponds to the decision direct approach which depends on the parameter β. The typical value of β is 0.98. The Wiener filter is the chosen spectral gain in the implementation and gain is given by

$$G_{DD}(f,k) = \frac{\widehat{SNR}_{Prio}^{DD}(f,k)}{1 + \widehat{SNR}_{Prio}^{DD}(f,k)} \qquad (4.25)$$

4.2.1 Two Step Noise Reduction

To improve the performance of the noise reduction algorithm, in two step noise reduction (TSNR) method, a priori SNR will be estimated in a two-step procedure. Following are the two steps for implementation:

Step 1:   Computation of the spectral gain $G_{DD}(f,k)$ using DD algorithm.
Step 2:   Estimation of a priori SNR at frame $f+1$ using the gain which is calculated from DD algorithm.

$$\widehat{SNR}_{Prio}^{TSNR}(f,k) = \widehat{SNR}_{Prio}^{DD}(f+1,k) \qquad (4.26)$$

$$= \beta' \frac{|G_{DD}(f,k)X(f,k)|^2}{\widehat{\gamma}_r(f,k)}$$

$$+ (1-\beta')D\left[\widehat{SNR}_{Post}(f+1,k) - 1\right] \qquad (4.27)$$

where β' plays the similar role as β but can have different value. To avoid the additional processing delay required to compute $\widehat{SNR}_{Post}(f+1,k)$ on the future frame $X(f+1,k)$, and to reduce the musical noise introduced with DD algorithm, set $\beta' = 1$. Then the above equation is degenerated as,

$$\widehat{SNR}_{Prio}^{TSNR}(f,k) = \frac{|G_{DD}(f,k)X(f,k)|^2}{\widehat{\gamma}_r(f,k)} \qquad (4.28)$$

Finally, the spectral gain will be computed as

$$G_{TSNR}(f,k) = h\left(\widehat{SNR}_{Prio}^{TSNR}(f,k), \widehat{SNR}_{Post}(f,k)\right) \qquad (4.29)$$

The above equation is used to enhance the noisy signal.

$$\widehat{S}(f,k) = G_{TSNR}(f,k)X(f,k) \qquad (4.30)$$

$$G_{TSNR}(f,k) = \frac{\widehat{SNR}_{Prio}^{TSNR}(f,k)}{1 + \widehat{SNR}_{Prio}^{TSNR}(f,k)} \tag{4.31}$$

TSNR algorithm can enhance the noise reduction performance since the gain matches to the current frame irrespective of the SNR.

4.2.2 Harmonic Regeneration Noise Reduction

The output of the TSNR still suffer from the harmonic distortion since some harmonics can be considered as noise only components that have been suppressed (Plapous et al. 2005). To overcome this harmonic distortion, harmonic regeneration noise reduction (HRNR) is proposed to create a full harmonic signal. In this method, a nonlinear function is applied to the time signal obtained in the process of TSNR. The restored signal is defined as

$$S_{harmonic}(t) = \varnothing\left(\widehat{S}(t)\right) \tag{4.32}$$

where $\varnothing$ is a nonlinear function.

$$\widehat{SNR}_{Prio}^{HRNR}(f,k) = \frac{\rho(f,k)\left|\widehat{S}(f,k)\right|^2 + (1 - \rho(f,k))S_{harmonic}(f,k)^2}{\widehat{\gamma}_r(f,k)} \tag{4.33}$$

The parameter $\rho(f,k)$ is used to control the mixing level of $\left|\widehat{S}(f,k)\right|^2$ and $|S_{harmonic}(f,k)|^2$.

$$G_{HRNR}(f,k) = \frac{\widehat{SNR}_{Prio}^{HRNR}(f,k)}{1 + \widehat{SNR}_{Prio}^{HRNR}(f,k)} \tag{4.34}$$

4.3 Speech Enhancement Based on DFrCT and W-HRNR

The block diagram of DFrCT-W-HRNR is shown in Fig. 4.2. First the noisy speech is segmented into small frames using window function and then the windowed samples are transformed into discrete fractional cosine transform with an appropriate order α.

General additive noise model is represented by

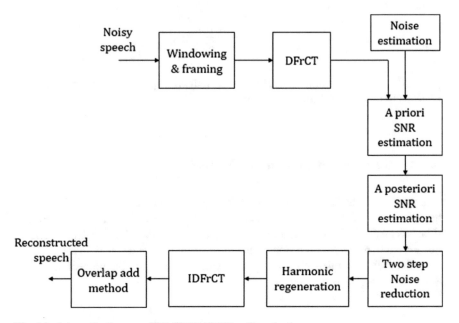

Fig. 4.2 Schematic diagram of DFrCT-W-HRNR noise reduction

$$p(t) = q(t) + r(t)$$

where $q(t)$ is the original /clean signal, $r(t)$ is the noise signal, and $p(t)$ is the noisy signal. Applying DFrCT we get,

$$P_C^\alpha(t) = Q_C^\alpha(t) + R_C^\alpha(t)$$

where $P_C^\alpha(t), R_C^\alpha(t)$, and $P_C^\alpha(t)$ are the DFrCT's original signal, noise signal, and noisy signal, respectively. The posteriori SNR, a priori SNR, and the instantaneous SNR is calculated in DFrCT domain using the following equations:

$$\text{SNR}_{\text{Post}C}(f,k) = \frac{\left| P_C^\alpha(f,k) \right|^2}{E\left[\left| R_C^\alpha(f,k) \right|^2 \right]} \tag{4.35}$$

and

$$\text{SNR}_{\text{Prio}C}(f,k) = \frac{E\left[\| Q(f,k) \| \right]^2}{E\left[\left| R(f,k) \right|^2 \right]} \tag{4.36}$$

$$\text{SNR}_{\text{Inst}C}(f,k) = \text{SNR}_{\text{Post}C}(f,k) - 1 \tag{4.37}$$

Spectral gain $G_C(f,k)$ in DFrCT domain is obtained by the function

$$G_C(f,k) = g\left(\widehat{\text{SNR}}_{\text{Prio}\,C}(f,k), \widehat{\text{SNR}}_{\text{Post}\,C}(f,k)\right) \tag{4.38}$$

$$\widehat{\text{SNR}}_{\text{Post}\,C}(f,k) = \frac{\left|P_C^\alpha(f,k)\right|^2}{\widehat{\gamma}_r(f,k)} \tag{4.39}$$

and

the SNR of decision direct approach in DFrCT domain is given as

$$\widehat{\text{SNR}}_{\text{Prio}\,C}^{\text{DD}}(f,k) = \beta \frac{\left|\widehat{Q_C^\alpha}(f-1.k)\right|^2}{\widehat{\gamma_{rc}}(f,k)} + (1-\beta)D\left[\widehat{\text{SNR}}_{\text{Post}\,C}(f,k) - 1\right] \tag{4.40}$$

The spectral gain of two step noise reduction $G_{\text{TSNR}_C}(f,k)$ is given as

$$G_{\text{TSNR}\,C}(f,k) = h\left(\widehat{\text{SNR}}_{\text{Prio}}^{\text{TSNR}}{}_C(f,k), \widehat{\text{SNR}}_{\text{Post}\,C}(f,k)\right) \tag{4.41}$$

The estimation of clean speech signal $\widehat{S}_C(f,k)$ in DFrCT domain is

$$\widehat{S}_C(f,k) = G_{\text{TSNR}\,C}(f,k)X(f,k) \tag{4.42}$$

The spectral gain of the harmonic regeneration noise reduction $G_{\text{HRNR}_C}(f,k)$ is defined as

$$G_{\text{HRNR}\,C}(f,k) = \frac{\widehat{\text{SNR}}_{\text{Prio}}^{\text{HRNR}}{}_C(f,k)}{1 + \widehat{\text{SNR}}_{\text{Prio}}^{\text{HRNR}}{}_C(f,k)} \tag{4.43}$$

The time domain output of the enhancement system can be obtained by taking inverse DFrCT of the Eq. (4.42).

The steps in the algorithm of DFrCT-W-HRNR are as follows:

Step 1:   Apply window function and frame the input noisy speech signal.
Step 2:   Convert the input signal into DFrCT domain.
Step 3:   Compute the noise statistics in DFrCT domain.
Step 4:   Calculate a priori SNR using decision direct approach.
Step 5:   Calculate gain of two step noise reduction method using Eq. (4.41).
Step 6:   Estimate the clean speech spectrum using TSNR gain.
Step 7:   Perform HRNR using Eq. (4.43).
Step 8:   The output of the Wiener filter is converted into time domain using inverse DFrCT and frame reconstruction using overlap and add method.

4.4   Performance Evaluation

To demonstrate the effectiveness of W-HRNR-DFrCT algorithm, the noisy speech signals sampled at 8 KHz are randomly taken from NOIZEUS speech corpus. The frame size was 256 samples with 50% overlap. The simulation is executed in MATLAB. In addition, enhanced signals were generated with the unconstrained, iterative Wiener filtering approach (Plapous et al. 2005). The evaluation of the algorithms is carried out by computing the values of SNR, segmental SNR (SNR$_{Seg}$), and SNR-loss, in five noisy conditions (airport noise, babble noise, train station noise, airport noise, and street noise) at three different SNR levels (0, 5, 10 dB). The parameters chosen for the simulation of W-HRNR and DFrCT-W-HRNR are shown in Table 4.2.

The objective measures SNR and SNR-loss are explained in Chap. 3. The method of computation for segmental SNR (*SNR$_I$*) is given below:

Time domain-based segmental SNR is one of the widely used objective measures to evaluate the performance of speech enhancement algorithms, which is formed by averaging the frame level of SNR estimate as given by

$$\text{SNR}_{\text{Seg}} = \frac{10}{M} \sum_{m=0}^{M-1} \log_{10} \frac{||s(m)||^2}{||s(m) - \widehat{s}(m)||^2}$$

where M denotes the number of frames, while $\widehat{s}(m)$ and $s(m)$ are the estimated and original speech vectors, respectively, in time domain. The segmental SNR values are limited in the range of $[-10, 35]$ dB in order to exclude frames with no speech.

The performance of DFrCT-W-HRNR method is first illustrated with time domain waveforms as spectrograms for qualitative analysis. In Fig. 4.3, time domain waveforms are presented for comparison between the Fig. 4.3a clean speech signal, Fig. 4.3b the noisy speech signal under 10 dB car noise condition, Fig. 4.3c output enhanced signals of the W-HRNR, and Fig. 4.3d the signal enhanced by DFrCT-W-HRNR. From Fig. 4.3d, it can be clearly seen that the DFrCT-W-HRNR algorithm enhanced the signal with improved SNR over the W-HRNR technique. The spectrograms of the signals are compared for the noise reduction performance in Fig. 4.4. From this figure, it can be noticed that DFrCT-W-HRNR better reduces the high frequency as well as low frequency noise components when compared to W-HRNR algorithm.

Table 4.2 Parameters chosen for simulation

Parameter	Value
Frame length	20 ms
Window function	Hanning
Overlap	50%
β, β'	0.98, 1
Size of FFT	512

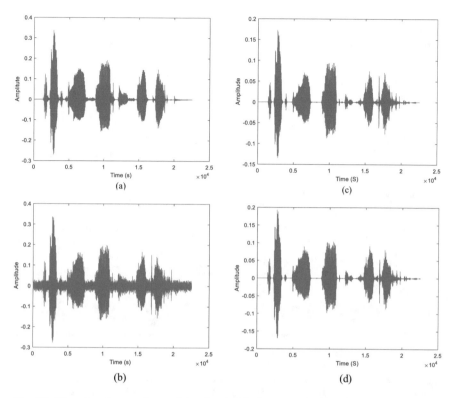

Fig. 4.3 Time domain waveforms of the signals (**a**) clean speech, (**b**) noisy speech (Car noise added at 10dB SNR), (**c**) signal enhanced by W-HRNR, and (**d**) signal enhanced by DFrCT-W-HRNR

Moreover, the results show that the proposed DFrCT-W-HRNR achieved better noise reduction as it outperforms the conventional W-HRNR method in terms of SNR for all noise types and SNR conditions as depicted in Table 4.3.

Besides the SNR, another objective measure called SNR_{Seg} is also compared for the DFrCT-W-HRNR and W-HRNR algorithms as shown in Table 4.4. Results affirmed the improved performance of the proposed DFrCT-W-HRNR over the W-HRNR method for all the five noise types under three noise levels.

Table 4.5 presents the objective scores of the intelligibility measure SNR_{Loss} for both the algorithms. From this table it is noted that DFrCT-W-HRNR achieved superior performance than the traditional W-HRNR in STFT domain.

4.5   Conclusions

- This chapter presents a single channel speech enhancement method by combining the fractional cosine transform and Wiener filter with harmonic regeneration noise reduction technique. DFrCT is a generalization of discrete cosine transform

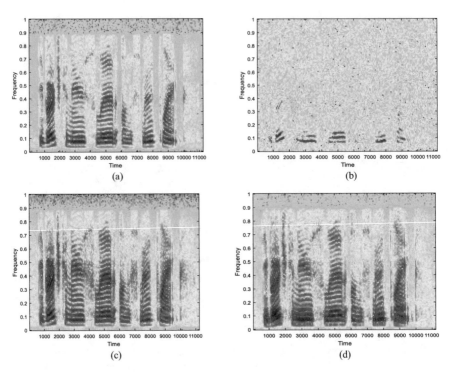

Fig. 4.4 Spectrograms of the (**a**) clean speech signal, (**b**) noisy speech signal (noise added at 0 dB), (**c**) signal enhanced by W-HRNR, and (**d**) signal enhanced by DFrCT-W-HRNR filter

Table 4.3 Comparison of SNR objective measure

Noise type	Input SNR (dB)	SNR (dB)	
		W-HRNR	DFrCT-W-HRNR
Babble	0	2.4848	3.0262
	5	4.3812	5.4013
	10	5.2620	6.5151
Car	0	3.4043	4.3227
	5	4.6752	5.9371
	10	4.7904	6.7965
Street	0	3.5162	4.1508
	5	4.6712	5.7187
	10	4.3525	6.2264
Airport	0	3.2928	4.0987
	5	4.4613	5.8182
	10	5.2496	6.9588
Train	0	3.8554	4.4947
	5	4.7388	5.9842
	10	5.2180	6.3145

Table 4.4 Comparison of segmental-SNR objective measure

Noise type	Input SNR (dB)	SNR_{Seg} (dB)	
		W-HRNR	DFrCT-W-HRNR
Babble	0	−0.2110	0.2080
	5	0.3017	0.8144
	10	0.8114	1.4097
Car	0	1.0054	1.3294
	5	1.7064	2.1606
	10	5.3504	1.4019
Street	0	0.3200	0.3495
	5	−0.9862	−0.4994
	10	1.3645	2.1152
Airport	0	−1.0286	−0.3938
	5	0.1091	0.6707
	10	1.5647	2.5067
Train	0	−0.2780	−0.1210
	5	0.2644	0.6266
	10	2.1076	2.2375

Table 4.5 Comparison of SNR_{Loss} objective measure

Noise type	Input SNR (dB)	SNR_{Loss} (dB)	
		W-HRNR	DFrCT-W-HRNR
Babble	0	0.9857	0.9841
	5	0.9674	0.9711
	10	0.9750	0.9783
Car	0	0.9945	0.9840
	5	0.9938	0.9915
	10	0.9827	0.9824
Street	0	0.9892	0.9745
	5	0.9497	0.9495
	10	0.9931	0.9883
Airport	0	0.9747	0.9734
	5	0.9717	0.9712
	10	0.9854	0.9800
Train	0	0.9713	0.9703
	5	0.9699	0.9692
	10	0.9984	0.9956

which is developed based on the eigenvectors of Hermitian matrix in the similar way of DFrFT. Because of using both the advantages in noise reduction of the fractional cosine domain and harmonic regeneration technique and the stable characteristic of the Wiener filtering technology, the proposed single channel speech enhancement method has achieved better performance by effectively removing the background noise from speech signals.

- The DFrCT algorithm obtained improvement in SNR by increasing the signal magnitude of the enhanced signal over conventional Wiener filter. This could be possible because of the additional degree of freedom offered with DFrCT by an angle of rotation in time frequency plane to match the signal characteristics better to reduce the noise from speech.
- Simulation examples proved that under different background noises and different noise levels, the DFrCT-W-HRNR speech enhancement method can obtain better speech recovery results than the traditional single channel enhancement method using Wiener filter with HRNR.

References

Bhagyashri Pandurangi, R., Patil, M. R., & Pujari, A. M. (2017). A Color image encryption technique using reality preserving fractional DCT. In *2017 2nd International Conference for Convergence in Technology, I2CT 2017, 2017-Janua* (pp. 832–836). https://doi.org/10.1109/I2CT.2017.8226245.

Bultheel, A., & Martínez-Sulbaran, H. (2006). Recent developments in the theory of the fractional Fourier and linear canonical transforms. *Bull Belg Math Soc Simon Stevin, 13*(5), 971–1005.

Chen, Y., Clamente, C., Soraghan, J., & Weiss, S. (2015). Fractional cosine transform (FRCT)-turbo based OFDM for underwater acoustic communication. In *2015 Sensor Signal Processing for Defence, SSPD 2015*, (Ici). https://doi.org/10.1109/SSPD.2015.7288503

Ko, L. T., Chen, J. E., Shieh, Y. S., & Sung, T. Y. (2012). A novel fractional discrete cosine transform based reversible watermarking for biomedical image applications. In *Proceedings - 2012 International Symposium on Computer, Consumer and Control, IS3C* 2012, (4), 36–39. https://doi.org/10.1109/IS3C.2012.19

Pei, S. C., & Yeh, M. H. (2001). The discrete fractional cosine and sine transforms. *IEEE Transactions on Signal Processing, 49*(6), 1198–1207. https://doi.org/10.1109/78.923302.

Plapous, C., Marro, C., & Scalart, P. (2005). Speech enhancement using harmonic regeneration. In *Proceedings. (ICASSP '05). IEEE International Conference on Acoustics, Speech, and Signal Processing, 2005* (Vol. 1, pp. 157–160). IEEE. https://doi.org/10.1109/ICASSP.2005.1415074

Thippeswamy, G., & Shekar, B. H. (2011). Face recognition based on discriminative fractional discrete cosine transform. In *Proceedings of the 5th Indian International Conference on Artificial Intelligence, IICAI 2011* (pp. 121–130).

Ya-ru, L., Guo-ping, L., & Jian-hua, W. (2015). Color image encryption algorithm with reality preserving fractional discrete cosine transform and spatiotemporal chaotic mapping. In *Proceedings of 2015 International Conference on Intelligent Computing and Internet of Things* (pp. 94–98). IEEE. https://doi.org/10.1109/ICAIOT.2015.7111546

Yoshimura, H. (2014). Fingerprint templates generated by the fractional fourier, cosine and sine transforms and their generation conditions. In *2014 World Congress on Internet Security, WorldCIS 2014* (pp. 30–34). https://doi.org/10.1109/WorldCIS.2014.7028161

Chapter 5
Fractional Sine Transform Based Single Channel Speech Enhancement Technique

5.1 Concepts of DST and DFrST

Discrete fractional sine transform (DFrST) is a generalized form of discrete sine transform (DST), which has an identical relationship with that between DFrFT and DFT. DFrST is introduced by Pei and Yeh (Pei and Yeh 2001). The definition of DFrST is expressed based on the Eigen decomposition of DST, similar to that of DFrFT derived from DFT. The computation of DFrFT for odd signals can be planted into half size DFrST similar to the way that even signals are planted into half size DFrST as discussed in Chap. 4.

The eigenvectors and eigenvalues of DFT kernel matrix are discussed in Chap. 2. DFrCT which is implemented by using the DFT Hermite eigenvectors is discussed in Chap. 4. In the similar manner DFrST is formulated by using the DFT Hermite eigenvectors and the DST kernel.

The DST has four types of kernel matrices as follows:

DST-I

$$S_{N+1}^I = \sqrt{\frac{2}{N}} \left[\sin\left(\frac{mn\pi}{N}\right) \right] \qquad (5.1)$$

for $m, n = 0, 1, \ldots N - 1$

DST-II

$$S_N^I = \sqrt{\frac{2}{N}} \left[k_m \sin\left(\frac{mn\pi}{N}\right) \right] \qquad (5.2)$$

for $m, n = 0, 1, \ldots N$

DST-III

© The Author(s), under exclusive licence to Springer Nature Switzerland AG 2020
P. Kunche, N. Manikanthababu, *Fractional Fourier Transform Techniques for Speech Enhancement*, SpringerBriefs in Speech Technology,
https://doi.org/10.1007/978-3-030-42746-7_5

$$S_N^I = \sqrt{\frac{2}{N}} \left[k_n \sin \left(\frac{mn\pi}{N} \right) \right] \tag{5.3}$$

for $m, n = 0, 1, \ldots N$
  DST-IV

$$S_{N+1}^I = \sqrt{\frac{2}{N}} \left[\sin \left(\frac{mn\pi}{N} \right) \right] \tag{5.4}$$

for $m, n = 0, 1, \ldots N$

where $k_m = \begin{cases} 1/\sqrt{2}, & m = 0 \text{ and } m = N \\ 1, & \text{others} \end{cases}$.

DST-I kernel has the properties of symmetry and periodicity. Hence DST-I is used for derivation of DFrST kernel matrices. The DST-I kernel which is chosen for DFrST is given as

$$S_N = \sqrt{\frac{2}{N-1}}$$

$$\times \begin{bmatrix} \sin \dfrac{\pi}{N+1} & \sin \dfrac{2\pi}{N+1} & \cdots & \cdots & \sin \dfrac{(N-1)\pi}{N+1} & \sin \dfrac{N\pi}{N+1} \\ \sin \dfrac{2\pi}{N+1} & \sin \dfrac{4\pi}{N+1} & \cdots & \cdots & \sin \dfrac{2(N-1)\pi}{N+1} & \sin \dfrac{2N\pi}{N+1} \\ \vdots & \vdots & \ddots & \cdots & \cdots & \vdots \\ \vdots & \vdots & \vdots & \ddots & \cdots & \vdots \\ \sin \dfrac{(N-1)\pi}{N+1} & \sin \dfrac{2(N-1)\pi}{N+1} & \vdots & \cdots & \sin \dfrac{(N-1)^2\pi}{N+1} & \sin \dfrac{N(N-1)\pi}{N+1} \\ \sin \dfrac{N\pi}{N+1} & \sin \dfrac{2N\pi}{N+1} & \vdots & \cdots & \sin \dfrac{N(N-1)\pi}{N+1} & \sin \dfrac{N^2\pi}{N+1} \end{bmatrix} \tag{5.5}$$

The N point DFrFT kernel is computed as

$$F_{N,\alpha} = V_N D_N^{2\alpha/\pi} V_N^T \tag{5.6}$$

$$= V_N \begin{bmatrix} 1 & & & & 0 \\ & e^{-j\alpha} & & & \\ & & \ddots & & \\ 0 & & & & e^{-j(N-1)\alpha} \end{bmatrix} V_N^T \tag{5.7}$$

Table 5.1 Eigenvalue multiplicities of the DST-I kernel matrices

N	Multiplicity of 1	Multiplicity of -1
Odd	$\frac{N+1}{2}$	$\frac{N-1}{2}$
Even	$\frac{N}{2}$	$\frac{N}{2}$

where $V_N = [v_0|v_1|\ldots\ldots.v_{N-1}]$, v_k is the kth order Hermite eigenvector of DFT, and α is the rotation angle. The DST-I eigenvectors can be attained from the DFT eigenvectors as shown below:

If $v = [0, v_1, v_2, \ldots\ldots\ldots.v_N, 0, -v_N, -v_{N-1}, \ldots\ldots -v_1]^T$ is an odd eigenvector of the $2(N+1)$-point DFT kernel matrix.

$$F_{2N+2}V = \lambda V$$

where $\lambda = j, -j$. Then

$$\tilde{V} = \sqrt{2}[v_1, v_2, \ldots\ldots\ldots v_N]^T \tag{5.8}$$

The above equation is the eigenvector of N-point DST-I kernel matrix where $j\lambda$ is the corresponding value.

$$S_N^I \tilde{V} = j\lambda\tilde{V} \tag{5.9}$$

The eigenvalues multiplicities of DST-I kernel matrices are given in Table 5.1. The N-point DFrST kernel is defined as

$$S_{N,\alpha} = \widehat{V_N}D_N^{2\alpha/\pi}\widehat{V_N^T} \tag{5.10}$$

$$= \widehat{V_N}\begin{bmatrix} 1 & & & & 0 \\ & e^{-2j\alpha} & & \\ & & \ddots & \\ 0 & & & e^{-j2(N-1)\alpha} \end{bmatrix}\widehat{V_N^T} \tag{5.11}$$

N-point DFrST kernel is computed in three steps:

1. Compute M_S−point DFT Hermite odd eigenvectors. $M_S = 2(N+1)$.
2. Compute the DST-I eigenvectors from the DFT Hermite even eigenvectors given in Eq. (5.8).
3. Determine the DFrST transform kernel using the following equation:

$$S_{N,\alpha} = \widehat{V_N} D_N^{2\alpha/_\pi} \widehat{V_N^T} \tag{5.12}$$

where $\widehat{V_N} = [v_1, v_2, \ldots \ldots . v_N]^T$ is the DST-I eigenvector obtained from kth order DFT Hermite eigenvectors.

The computation of DFrFT, DFRCT, and DFrST would take $O(N^2)$ complexity with the ordinary multiplications.

5.1.1 Properties of DFrST

1. Unitary: The DFrST is unitary.

$$S_{N,\alpha}^* = S_{N,\alpha}^{-1} = S_{N,-\alpha} \tag{5.13}$$

2. Angle additive: The DFrST satisfies the property of angle additivity similar to DFrFT.

$$S_{N,\alpha} \, S_{N,\beta} = S_{N,\alpha+\beta} \tag{5.14}$$

3. Periodic: The DFrFT is periodic with period 2π whereas DFrST is periodic with period π.

$$S_{N,\alpha+\pi} = S_{N,\alpha} \tag{5.15}$$

4. Symmetric: The DFrST kernel holds the symmetry property.

$$S_{N,\alpha}(a, b) = S_{N,\alpha}(b, a) \tag{5.16}$$

5.1.2 Applications of DFrST

Hiroyuki (Yoshimura 2014) proposed new finger print templates based on one dimensional discrete fractional Fourier, cosine, and sine transforms to resize the fingerprint recognition system with high accuracy and high robustness against attacks. Results demonstrated that finger prints generated by DFrCT and DFrST are more appropriate than DFrFT. Bharati et al. (Salunke and Salunke 2016) analyzed and compared the encrypted images using DFrFT, DFrST, and DFrCT.

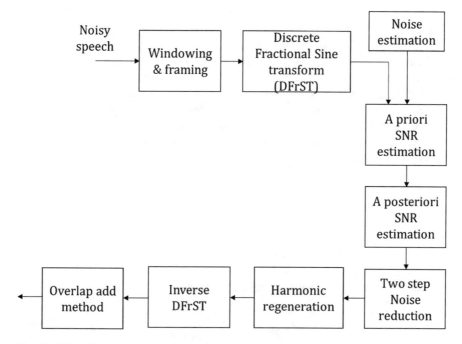

Fig. 5.1   Block diagram of proposed speech enhancement system

5.2   Speech Enhancement Based on DFrST

The schematic diagram of proposed algorithm for speech enhancement is shown in Fig. 5.1. At first, the noisy speech signal is segmented into frames by using windowing and framing. Then the discrete fractional sine transform is applied to the frames of the signal to transform the signal into p order DFrST domain. Noise spectral estimation is computed by taking the initial silence period into account. Then, a priori SNR values and a posteriori SNR values are computed. The VAD decision is subsequently computed for each frequency bin based on the resulting a priori SNR values. Finally, the clean speech spectral amplitude is estimated. In the computation of the a priori SNR the DD bias in low SNR conditions is compensated using an alternative approach, the bias could be reduced to compute the a priori SNR.

A speech signal x is contaminated by additive noise n so that the distorted speech is expressed as

$$y = x + n \tag{5.17}$$

This system can be modeled in the discrete fractional sine transform domain as

$$Y(k, i) = X(k, i) + N(k, i) \tag{5.18}$$

where $Y(k, i)$, $X(k, i)$, and $N(k, i)$ represent the kth spectral component of the short time frame i of the DFrST transform of the speech elements y, x, and n, respectively. The frame index i is dropped wherever possible for better readability, thus, $S_y(k)$, $S_x(k)$, and $S_n(k)$ are the power spectra of noisy speech which underlies clean speech and input additive noise, respectively. The noise reduction process consists in the application of a spectral gain $G(k)$ to each short time DFrST spectrum value $Y(k)$. The spectral gain is calculated based on two parameters: the a posteriori signal-to-noise ratio and the a priori signal-to-noise ratio (SNR).

The enhanced speech can be obtained by applying the gain function to the noisy speech as follows:

$$\widehat{X}(k) = G(k)Y(k) \tag{5.19}$$

where

$$G(k) = g(\widehat{\gamma}(k), \widehat{\varepsilon}_{DD}(k)) \tag{5.20}$$

The gain function g for Weiner filter is defined as

$$G_{\text{weiner}}(k) = \frac{\widehat{\varepsilon}_{DD}(k)}{1 + \widehat{\varepsilon}_{DD}(k)}, \tag{5.21}$$

To improve the noise reduction performance of algorithm, two step noise reduction (TSNR) is used.

The $G_{\text{weiner}}(k, i)$ spectral gain is determined as in Eq. (5.21) in the first step of the TSNR and is used to produce an initial guess for the enhanced expression

$$G_{\text{TSNR}}(f, k) = \frac{\widehat{\varepsilon}_{\text{Prio}}^{\text{TSNR}}(k, i)}{1 + \widehat{\varepsilon}_{\text{Prio}}^{\text{TSNR}}(k, i)} \tag{5.22}$$

$$\widehat{\varepsilon}_{\text{Prio}}^{\text{TSNR}}(f, k) = \frac{|G_{\text{weiner}}(k, i)X(k, i)|^2}{\widehat{\gamma}(k, i)} \tag{5.23}$$

In this method, a nonlinear function is applied to the time signal obtained in the process of TSNR. The restored signal using harmonic regeneration noise reduction (HRNR) is defined as

$$S_{\text{harmonic}}(t) = \varnothing \left(\widehat{S}(t) \right) \tag{5.24}$$

where $\varnothing$ is a nonlinear function.

$$\widehat{\varepsilon}_{\text{Prio}}^{\text{HRNR}}(k,i) = \frac{\rho(k,i)\left|\widehat{S}(k,i)\right|^2 + (1 - \rho(k,i))S_{\text{harmonic}}(k,i)^2}{\widehat{\gamma}_r(k,i)} \tag{5.25}$$

The parameter $\rho(k,i)$ is used to control the mixing level of $\left|\widehat{S}(k,i)\right|^2$ and $|S_{\text{harmonic}}(k,i)|^2$.

$$G_{\text{HRNR}}(k,i) = \frac{\widehat{SNR}_{\text{Prio}}^{\text{HRNR}}(k,i)}{1 + \widehat{SNR}_{\text{Prio}}^{\text{HRNR}}(k,i)} \tag{5.26}$$

This TSNR output will have some distortion and to overcome that HRNR method is employed.

The gain of Weiner filter based on HRNR method in DFrST domain is given in Eq. (5.26)

The proposed algorithm for speech enhancement is described as follows:

Step 1: Apply window function and frame the input signal.
Step 2: Convert the input noisy speech signal into DFrST domain.
Step 3: Estimate the noise spectrum in DFrST.
Step 4: Linear Wiener filtering with two step noise reduction and harmonic regeneration noise reduction techniques.
Step 5: The output of the Wiener filter is converted into time domain using inverse DFrST and frame reconstruction using overlap and add method.

5.3 Performance Evaluation

The noisy speech signals used in this experiment are taken from the NOIZEUS ("NOIZEUS: A noisy speech corpus for evaluation of speech enhancement algorithms" n.d.) database, which has been developed to make easier for the research groups to compare speech enhancement algorithms. This database consists of a total of 30 IEEE sentences spoken by female and male speakers. Each speech signal in the database is recorded using a 16-kHz sampling rate. The noise was taken from the AURORA database and includes suburban train noise, babble, car, exhibition hall, restaurant, street, airport, and train station noise. In this experiment, we take random sentences from database for evaluation. To demonstrate the effectiveness of W-HRNR-DFrST algorithm, the simulation is conducted in MATLAB. The signals are enhanced using the iterative Wiener filtering approach that is presented in Sec. 5.2.

The evaluation of the algorithms is carried out by computing the values of SNR, SNR_{LOSS}, and segmental-SNR under five noisy conditions (airport noise, babble noise, car noise, train station noise, and street noise) at different SNR levels (0, 5, 10 dB). Then the average values are computed for each noise type and compared with speech enhancement algorithms like Wiener filter with harmonic regeneration noise reduction and discrete fractional cosine transform based W-HRNR (DFrCT-W-HRNR) and discrete fractional sine transform based W-HRNR (DFrST-W-HRNR).

5.4   Results and Observations

Figure 5.2 shows the comparison for the time domain waveforms of the speech signals. Figure 5.2a–d shows the original signal, signal processed by proposed DFrST-W-HRNR, signal processed by DFrCT-W-HRNR, and signal processed by the DFrFT-W-HRNR. Comparing Fig. 5.2b–d, it can be noticed that the proposed method gives a cleaner result than other two approaches. The comparison of the spectrograms of the clean speech, enhanced signals by DFrST, DFrCT, and DFrFT are presented in Fig. 5.3. From these figures it can be inferred that DFrST based W-HRNR algorithm reduces noise better than the DFrCT and DFrFT based Wiener algorithms. In Fig. 5.4, the comparison of power spectral densities of clean speech, DFrST, DFrCT, and DFrFT based W-HRNR algorithms is provided. From the two figures it can be clearly observed that DFrST-W-HRNR gives better performance in terms of signal estimation.

Moreover, the results show that the proposed DFrST-W-HRNR achieved better noise reduction as it outperforms the DFrCT-W-HRNR and the conventional W-HRNR methods in terms of SNR for all noise types and SNR conditions as

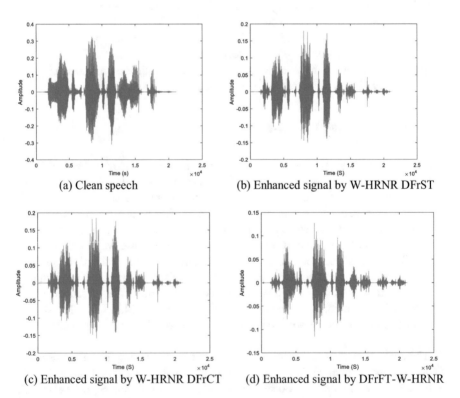

(a) Clean speech   (b) Enhanced signal by W-HRNR DFrST

(c) Enhanced signal by W-HRNR DFrCT   (d) Enhanced signal by DFrFT-W-HRNR

Fig. 5.2 Time domain waveforms of the signals (**a**) clean speech, (**b**) noisy speech, (**c**) signal enhanced by NLMs-FrFT, and (**d**) signal enhanced by NLMS

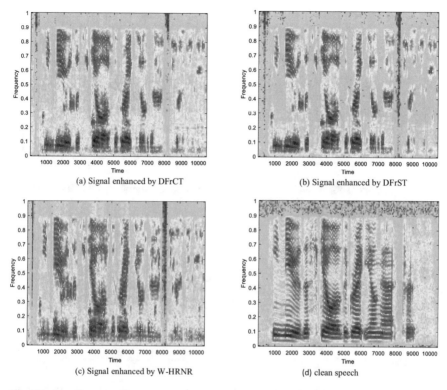

(a) Signal enhanced by DFrCT

(b) Signal enhanced by DFrST

(c) Signal enhanced by W-HRNR

(d) clean speech

Fig. 5.3 Spectrograms of the signal enhanced by (**a**) DFrCT-W-HRNR (noise added at 0 dB), (**b**) DFrST-W-HRNR (**c**) W-HRNR, and (**d**) clean speech

depicted in Table 5.2. Besides the SNR, another objective measure called SNR-Seg of DFrST-W-HRNR is also compared with the DFrCT-W-HRNR and W-HRNR algorithms as shown in Table 5.3. Results affirmed the improved performance of the proposed DFrST-W-HRNR over the DFrCT-W-HRNR and classical W-HRNR methods for all the five noise types under three noise levels. Table 5.4 presents the scores of the intelligibility measure SNR_{Loss} for all the three algorithms (DFrST-W-HRNR, DFrCT-W-HRNR and W-HRNR). From this table it is noted that DFrST-W-HRNR achieved superior performance than the traditional W-HRNR in STFT domain and DFrCT based W-HRNR algorithms.

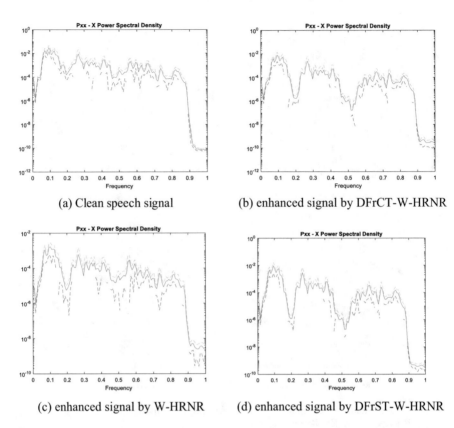

(a) Clean speech signal

(b) enhanced signal by DFrCT-W-HRNR

(c) enhanced signal by W-HRNR

(d) enhanced signal by DFrST-W-HRNR

Fig. 5.4 Power spectral densities of the (**a**) clean speech signal, signal enhanced by (**b**) DFrCT-W-HRNR (noise added at 0 dB), (**c**) W-HRNR, and (**d**) DFrST-W-HRNR filter

Table 5.2 Comparison of SNR objective measure

Noise type	Input SNR (dB)	SNR (dB)		
		W-HRNR	DFrCT-W-HRNR	DFrST-W-HRNR
Babble	0	2.4848	3.0262	3.4326
	5	4.3812	5.4013	5.7042
	10	5.2620	6.5151	6.8782
Car	0	3.4043	4.3227	4.9582
	5	4.6752	5.9371	6.5453
	10	4.7904	6.7965	7.2635
Street	0	3.5162	4.1508	4.7567
	5	4.6712	5.7187	5.8799
	10	4.3525	6.2264	6.7028
Airport	0	3.2928	4.0987	4.2883
	5	4.4613	5.8182	6.1949
	10	5.2496	6.9588	7.4469
Train	0	3.8554	4.4947	4.9292
	5	4.7388	5.9842	6.4493
	10	5.2180	6.3145	6.9847

Table 5.3 Comparison of segmental-SNR objective measure

Noise type	Input SNR (dB)	SNR$_{Seg}$ (dB)		
		W-HRNR	DFrCT-W-HRNR	DFrST-W-HRNR
Babble	0	−0.2110	0.2080	0.4697
	5	0.3017	0.8144	1.0032
	10	0.8114	1.4097	1.5832
Car	0	1.0054	1.3294	1.6128
	5	1.7064	2.1606	2.7581
	10	5.3504	1.4019	1.7127
Street	0	0.3200	0.3495	0.7355
	5	−0.9862	−0.4994	−0.4629
	10	1.3645	2.1152	2.4427
Airport	0	−1.0286	−0.3938	−0.1248
	5	0.1091	0.6707	0.7788
	10	1.5647	2.5067	2.7284
Train	0	−0.2780	−0.1210	0.1396
	5	0.2644	0.6266	0.7533
	10	2.1076	2.2375	3.0066

Table 5.4 Comparison of SNR$_{Loss}$ objective measure

Noise type	Input SNR (dB)	SNR$_{Loss}$ (dB)		
		W-HRNR	DFrCT-W-HRNR	DFrST-W-HRNR
Babble	0	0.9857	0.9841	0.9792
	5	0.9674	0.9711	0.9602
	10	0.9750	0.9783	0.9717
Car	0	0.9945	0.9840	0.9813
	5	0.9938	0.9915	0.9826
	10	0.9827	0.9824	0.9732
Street	0	0.9892	0.9745	0.9721
	5	0.9497	0.9495	0.9380
	10	0.9931	0.9883	0.9834
Airport	0	0.9747	0.9734	0.9656
	5	0.9717	0.9712	0.9618
	10	0.9854	0.9800	0.9729
Train	0	0.9713	0.9703	0.9604
	5	0.9699	0.9692	0.9586
	10	0.9956	0.9984	0.9926

5.5 Conclusions

- This chapter presents a single channel speech enhancement method by combining the fractional cosine transform and Wiener filter with harmonic regeneration noise reduction technique. DFrST is a generalization of discrete sine transform which is developed based on the eigenvectors of Hermitian matrix in the similar way of DFrFT. In this method the noisy speech signal is transformed into discrete fractional sine transform domain to an order p.

- Simulation examples confirm that under different colored noise, the proposed single channel speech enhancement method can obtain better speech recovery results than the traditional Wiener filter speech enhancement method and fractional cosine transform based speech enhancement method.

References

NOIZEUS: A noisy speech corpus for evaluation of speech enhancement algorithms. (n.d.). Retrieved from https://ecs.utdallas.edu/loizou/speech/noizeus/

Pei, S. C., & Yeh, M. H. (2001). The discrete fractional cosine and sine transforms. *IEEE Transactions on Signal Processing, 49*(6), 1198–1207. https://doi.org/10.1109/78.923302.

Salunke, B. A., & Salunke, S. (2016). Analysis of encrypted images using discrete fractional transforms viz. DFrFT, DFrST and DFrCT. In *International Conference on Communication and Signal Processing, ICCSP 2016* (pp. 1425–1429). https://doi.org/10.1109/ICCSP.2016.7754390

Yoshimura, H. (2014). Fingerprint templates generated by the fractional fourier, cosine and sine transforms and their generation conditions. In *2014 World Congress on Internet Security, WorldCIS 2014* (pp. 30–34). https://doi.org/10.1109/WorldCIS.2014.7028161

Chapter 6
Summary and Perspectives

6.1 Summary

The work resented in this book is based on the background noise reduction of speech signal in fractional Fourier domain (FrFD). In FrFD, the signal will be transformed to any intermediate domain in between time and frequency planes. Some additional gain in the performance can be obtained with the fractional order parameter used in FrFT compared to time or frequency domain techniques. In particular, for filtering in FrFD, the MSE score could be reduced more over the traditional filtering approaches. FrFT is a potent signal processing tool for the speech signal since it works well in the estimation of parameters and the FrFD filtering.

FrFT has been applied in many of the signal processing applications so far. Recently, FrFT has also gained the attention of speech processing community. But, very few works reported the speech enhancement based on FrFT. To bridge this gap, an attempt has been made in this book, to review the existing FrFT-based enhancement methods and also study its performance in both the single and dual channel enhancement systems. Moreover, DFrCT and DFrST have also been investigated for single channel enhancement.

The study of the FrFT-based speech enhancement begins with the investigation of the signal filtering in fractional Fourier domain in dual channel adaptive noise cancellation which has been presented in Chap. 3. To study the different noise effects of denoising at different noise levels additive white Gaussian noise (AWGN) is used.

The advancement of DFrFT and its increased applications supported the development of different class of fragmentary changes like fractional cosine, sine, Hadamard, and Hartley transform. The DFrCT and DFrST are the generalization of DCT and DST which are derived in a similar manner to the derivation of DFrFT from DFT eigenvectors. Only few studies have reported the use of fractional cosine and sine transforms for the signal and image processing applications so far and there is no literature available on the use of DFrCT and DFrST for speech processing

© The Author(s), under exclusive licence to Springer Nature Switzerland AG 2020   99
P. Kunche, N. Manikanthababu, *Fractional Fourier Transform Techniques*
for Speech Enhancement, SpringerBriefs in Speech Technology,
https://doi.org/10.1007/978-3-030-42746-7_6

especially, speech enhancement. In this book the comparison is carried out on three techniques that is DFrFT, DFrCT, and DFrST for the application of speech processing.

In Chap. 4, the DFrCT algorithm is employed for Wiener filter noise reduction approach. DFrCT achieved superior performance for single channel enhancement systems over the traditional Wiener filter and DFrFT based techniques. The proposed technique nearly offered 1–2 dB improvement in the SNR over W-HRNR method in short time Fourier transform (STFT) domain.

In Chap. 5, the signal filtering in discrete fractional sine transform domain is implemented in single channel noise cancellation system. DFrST outperformed both the DFrFT and DFrCT for single channel enhancement systems. The proposed DFrST-W-HRNR technique nearly offered 1–2 dB improvement in the SNR over DFrCT method, 2–3 dB improvement over DFrFT, and 3–4 dB improvement over Wiener filter in short time Fourier transform (STFT) domain.

6.2 Future Scope

- The study of the FrFT-based speech enhancement can be extended by deploying FrFT into some other enhancement techniques such as subspace enhancement methods.
- The order of the FrFT can be optimized by using optimization techniques to get improved optimal order that exactly matches with the signal characteristics.
- FrFT can also be combined with artificial intelligence and machine learning techniques, with the aim of developing a more efficient system for speech enhancement.

Index

Printed in the United States
By Bookmasters